A CONSERVATION FOUNDATION STUDY

PESTICIDES and
the LIVING LANDSCAPE

THE UNIVERSITY OF WISCONSIN PRESS, MADISON and MILWAUKEE

PESTICIDES and

the LIVING LANDSCAPE

ROBERT L. RUDD

Published 1964
THE UNIVERSITY OF WISCONSIN PRESS
Box 1379, Madison, Wisconsin 53701

Printings 1964, 1966, 1970

Printed in the United States of America

ISBN 0-299-03210-8 cloth, 0-299-03214-0 paper
LC 64-14506

CONTENTS

NOTE TO THE 1970 PRINTING

Six years have passed since the original printing of *Pesticides and the Living Landscape*. At this reprinting there seems to be no reason for major changes. The chief substance of the book stands now as then. Biological responses, precautions to observe, and suggestions for change apply now as then. The dilemma I describe in Chapter 1 remains with us. If anything, the dilemma has become more difficult to resolve.

But there are some changes in the subject area worth noting. Some concern actual shifts in pest-control practices; others, the elucidation of the mechanisms of widespread pesticide-residue transport and resulting biological effects. Yet others — most important — concern changes in attitude toward problems caused by pesticides and other products of technological man.

The first changes lie with the use of pesticide itself. There has been an accelerating shift to use of less persistent chemicals. The magnitude of this shift is not as great as I believe it should be. The essential universality of chemical residues argues for a more rapid rate of change away from the use of persistent chemicals. Yet the dilemma faces pest-control advisors too. There are no ready substitutes in many instances. The real problems of maintenance of health, of production of foods and fibers, and of economics remain. The immediacy of these demands coupled frequently with the lack of effective alternatives continues to dictate that pest-control practitioners will not quickly depart from their dependency on longer-lived chemicals. Indeed, these practitioners in general will not initiate such changes. Changes are more likely to come from administrative and legislative actions dictated by wider considerations. And, as I have indicated in the text, the less persistent chemicals have their own associated problems requiring resolution, the greatest of which is an ever-widening insecticide resistance. Such a response would be expected from expanded and repetitive use.

Optimistic forecasts for the general use of biological methods in pest control are not supportable. It is true that sex hormones as attractants or as reproductive inhibitors have been employed successfully in specific instances. It is also true that integrated pest control dependent on timing, natural counterforces, and minimal pesticide use has become more widely accepted. But it is more widely accepted as a desirability than a practicality. In major crops wherein widely distributed pest species remain the focus of control, pesticides continue to be the choice for use.

Although the shift in practice is slight and unsatisfactory, changes in attitude, among entomologists particularly, have led to major alterations in administration and research. For example, state and federal forestry officials recommend treatment with DDT only in extreme cases. The effect is a virtual ban on the use of DDT and other chlorinated hydrocarbons on public lands. Many states, provinces, and countries now have limited or total bans on DDT and some other pesticides. Many agencies and advisors have announced "phase-out" periods. As another example, the research and developmental efforts of the U.S. Department of Agriculture are now largely dedicated to biological aspects of pest control. Accordingly, the stage has been set for reducing contamination levels and for offering biological alternatives in pest control as discovery permits.

Ecologists, professional and amateur, have been responsible for both discovery and controversy regarding the effects of pesticides in the field. In a sense, these ecologists have constituted an informal monitoring system. From them have come many pressures for change in pest-control practice. That uncontrolled residues in ecosystems did in fact produce general toxic effects was recognized only gradually by many people concerned with the production and application of pesticides. Conviction is now largely achieved. It came most readily when the physiological and toxicological mechanisms by which these residues acted could be explained, or at least suggested. The studies of the Patuxent Research Refuge of the U.S. Fish and Wildlife Service are particularly significant.

Still, a continuing problem faced ecologists in the sixties. The description of effects could not explain how these effects are produced. The greatest single blessing of the last several years was that physiologists and molecular biologists joined the group of interested ecologists. The juncture has been termed chemical ecology or environmental chemistry. This fortunate circumstance has already contributed a great deal. I mention only two important examples. The field observation, first made in England, that residues in the eggs of raptorial birds are responsible for several behavioral alterations and ultimately for population declines has now been linked to upset calcium balances in laying birds. The mechanisms and the generality of these effects have recently been described by D. B. Peakall (*Scientific American,* 122: 72; 1970).

The second example concerns the entire principle of threshold levels of action on which tolerance limits are based. Clearly, residue levels that produce the "egg-shell thinning" effect are far below levels presumed to be safe by existing regulations. It was R. W. Risebrough (personal communication and in *Chemical Fall-out,* Charles C Thomas Co., 1969) who suggested that we may have to discard the entire "parts per million" approach. It was also primarily Risebrough who showed that non-pesticide chlorinated hydrocarbons (the polychlorinated biphenyls) were distributed in the earth's ecosystems in much the same way as pesticides. In short, the problem of environmental consequences of widely distributed toxic residues is larger than we have thought.

Though awareness that problems of general importance are created by pesticides came slowly to pest-control advisors, and even more slowly to others in professional and managerial disciplines impinging on pest control, in the last year or two the situation has altered dramatically. Environmental pollution is a common concern, ecology a household word. Recognition of the proper social role of the ecological specialist seems general. This attitude permeates the most recent summary of the effects and general significance of pesticides. The Report of the Secretary's Commission on Pesticides and their Relationship to Environmental Health (the so-called *Mrak Report;* U.S. Department of Health, Education and Welfare, 1969) logically enough identifies pesticide use as a far more general concern than any similar report has done in the past.

The most enlightened change in recent years, in my view, is the recognition that difficulty with pesticides is part of a much larger difficulty — that of human adjustment to an intensifying technology. The difficulty of living with the waste products of technology has now been recognized as one problem with many facets. Environmental pollutants from whatever source — automobiles, mist-blowers, sewers, industrial manufacture — can and do degrade the environment. The ultimate in such degradation is the loss of utilizable biological productivity. Lake Erie is not only a topical example. It is a symbol of the many "Lake Erie's" manifest in so many ways in our technological culture. Public awareness of this inadaptability of technological man is now the greatest in memory. As I approach participation in the newly instituted Ecology Week and the national Environmental Teach-In, I am pleased at the attention being given to the general problem. But such attention must be both effective and continuing. A great deal remains to be understood. Adapting to ourselves is the fundamental challenge. This book is one attempt to meet that challenge.

R. L. R.

Davis, California
April, 1970

PREFACE

No creature before man conceived of or achieved the ability to change the earth's surface according to its will. The land surface of the earth bears clear evidence that in some places, pock-marked and discontinuous, the dominance of man over environment is now established. Politically, we now label these pocks the well-developed, the favored, the modern nations. The spaces between them we call the backward, the underdeveloped, or—once popular—the "have-not" nations. It is the current fashion of the favored nations to set about applying to the larger remainder of the world the same methods that have allowed their rise to relative affluence. This commitment to total dominance over the producing portions of the earth, irrespective of political patterns, is a new event in earth history. Limited technology historically has confined the scope of man's domination and dulled his dreams of full mastery. Even so limited, local domination frequently had tragic results as so many magnificent ruins attest. The immediate successes of modern technology are not clouded by pessimistic reflec-

tions on the past. Rather, an optimistic view prevails. We no longer
need dream of a world in which material security and comfort are available to all. Seemingly we have the means to achieve them and we
should, as we are constantly persuaded, set ourselves with conviction
to the task of accomplishing the ends as quickly as possible. Only a few
voices ask what is the product to be and what is its price?

In our optimistic flush we may fail to see that both means and ends
of man's dominance require constant appraisal. This book is an attempt
to evaluate only one of many avenues to man's mastery of his environment. It rests on the question—How can we control the many thousands of plant and animal species that compete with us for food, fiber,
and timber or in some way threaten our health and comfort and at the
same time recognize, preserve, and enhance the productive, cultural,
and spiritual values that the living environment gives to us?

Therefore, this book treats relevant aspects of five subjects, not one:

(1) How man exploits the living environment.

(2) How he challenges the organisms that interfere with his exploitation.

(3) How successful he is in controlling these competing organisms.

(4) The methods he uses to control them.

(5) The price he pays for the methods in practice.

The last-listed is my chief emphasis.

None of the subjects is new, but there is a particularly good reason
why they should be examined now. Methods of controlling competing
organisms have become dependent on widespread application of poisons that, while controlling, produce undesirable effects. The effects of
control agents, when not limited to a particular pest species, immediately give rise to questions of value. How much will we sacrifice to
achieve the desired control of one organism? This question is currently
a heated one, with viewpoints ranging to the extremes. Somewhere between these extremes lies a compromise zone in which practical decisions lie. The likelihood of pleasing everyone on the question of
values is very small indeed. But, values aside, the suspicion arises in
the minds of many that current emphasis on chemical control agents
may be self-defeating in the long run. Moreover, the subtlety and magnitude of chemical effects are only now becoming fully apparent. Some
of these effects, although not immediately visible to all, have impressed
even the most tolerant that something is wrong with current chemical

control practices, and with the educational and governmental philosophies that support them.

There are, then, three classes of objection to present emphasis on the control of living things by chemicals:

(1) From those who deplore the change in the organic environment that inevitably accompanies chemical use.

(2) From those who feel that the balance of value judgments is awry—that the values received are not weighed equally with the losses.

(3) From those specialists who believe that biological complications resulting from chemicals are so great that chemical control is self-defeating, and who believe that alternative methods of control should be applied more widely.

This book describes some of the conflict between man and other living things, and the methods we have used to reduce this conflict. It is influenced throughout by the objections listed above, and may therefore seem to show inordinate preoccupation with real and potential hazards, with imbalanced values, and with alternative control procedures. This emphasis is intentional.

My single purpose throughout this book is to explain to the serious reader, particularly one with responsibilities relevant to the subject, what the kinds of hazards resulting from chemical pest control are. There are three levels of explanation. The most general is Chapter 1, in which I attempt a capsule description of the choices open to us. Chapters 2 through 5 are intended to be summary reviews of the kinds, economics, and regulation of pesticides—for the benefit of readers not familiar with their character. The remaining 17 chapters constitute the bulk of description and form the basis for my personal judgments and recommendations intercalated throughout.

R. L. R.

Davis, California
September, 1963

ACKNOWLEDGMENTS

In 1958 the officers of the Conservation Foundation suggested that I undertake a general study of the influence of pesticides in nature. At that time neither the Foundation nor I knew quite where such a study would lead. The Foundation generously provided me with encouragement, continuing counsel, and financial aid to lighten the task. I am grateful to all the Foundation, and particularly to Dr. Fairfield Osborn, Mr. Samuel Ordway, Dr. F. Fraser Darling, and Mr. Stephen Bergen. The Foundation's most generous gift, however, was complete freedom to write as I wished. Accordingly, the views expressed in this book are my own and do not necessarily agree with those held by the Foundation or its individual officers.

The substance of the book derives from hundreds of published works, countless discussions, and frequent correspondence with interested individuals in many countries. I cannot acknowledge by name all those people who assisted in some way. But the list is long. I say only that I am grateful.

I acknowledge with warm thanks my colleagues Professors M. A. Miller and G. W. Salt who read the manuscript for technical accuracy. I have to thank particularly Mr. Thompson Webb and the editorial staff of the University of Wisconsin Press for their continued patience and care.

I · PERSPECTIVE

1 · DISTURBANCE AND DILEMMA

Man's accomplishments in preventing disease, in providing himself with food and shelter, and in increasing the efficiency of his labors to permit himself more leisure time are impressive. Many are so impressed with his abilities to change the living landscape in a fashion best suited to his immediate interests that they believe man can become exempt from the biological laws which govern his life and the lives of perhaps two million species of plants and animals. Man has been termed an *ecological dominant*—a creature capable of molding his environment to his will, yet independent of it. Biblical allegory reinforces this belief that he was apart from, not a part of, the naturalistic world.

But his dominance is not so firmly based. The living fabric of the earth's surface is of delicate and complicated weave. Nature is quite as capable of preventing fulfilment of our aspirations if mistreated as of rewarding us if understandingly managed. Sustained dominance can come only through genuine understanding of the natural forces we have

3

set about to guide and of which we are an integral part. Concomitance —living with natural forces—rather than dominance is the only route to enduring self-interest.

The illusion of ecological dominance can be illustrated with a multitude of examples of environmental abuse—the barren lands induced by overgrazing of livestock; the soils worn away by inappropriate plantings; the once-forested watersheds that no longer store and restrain water; the streams, formerly clean and productive, that became the sewers of urban and industrial man; the threat to living beings from uncontrolled fall-out from nuclear explosions. But none of these examples illustrates any more clearly than the subject of this book— chemical pest control—the need for ecological concomitance.

The tools of pest control are now largely synthetic chemicals. These chemicals—pesticides—are designed to kill or in some way inhibit the plant and animal competitors that interfere with our health, comfort, or production of foods and fibers. Some 200 basic chemicals are commonly used in agriculture and these are commercially presented in thousands of different formulations under many trade names. These chemicals are broadly categorized, according to their intended targets, as insecticides, rodenticides, fungicides, and herbicides. Most can be considered "biocides"—they kill living things. The same or similar chemicals are used away from agriculture in public health, forestry, fish and game management, rights-of-way maintenance, property protection, and recreation.

Generally speaking, the purposes of pesticidal use are clear and widely accepted. We realize, for example, that successes in pest control have, along with other technological applications, greatly changed the yields in American forest, grazing, and crop environments. Indirectly, we may be assured that crop production, toward which empirical pest control has contributed, has risen greatly in the last five decades. The area now under agriculture, for example, is about the same as it was in 1910; but only one person in five now lives and works on a farm in contrast to the three out of five in 1910 (Clawson *et al.*, 1960). We may also take as a measure of the success of agricultural technology—in which pest control prominently figures—the ratio between numbers of farm workers and consumers. In 1920, one farm worker produced food and fiber for eight, whereas in 1957 the ratio was one to twenty-three (Gunther and Jeppson, 1960). Finally we can accept as common observation that agricultural products of high quality and great diversity are readily available to us all. In the basic crops at least, we can also note that surpluses necessitating federal support and storage schemes cost a great deal of money (some $8 billion annually) and political contention.

But if chemical pest control is indeed so valuable, why should anyone argue against it? Here are a few reasons:

(1) Most pesticides are nonselective; they kill forms of life other than their pest targets.

(2) Their manner of use, though often increasing selectivity, is in most cases not precise, the pesticide being restricted neither to the pest species nor to the area where applied.

(3) Insufficient attention has been given to alternative means of crop protection. Most alternative methods of pest control are culturally or biologically based and result in greater diversity and stability in the biota. Pesticides reduce faunal diversity—a natural control—and ensure instability.

(4) Many kinds of chemicals (notably the chlorinated hydrocarbon group to which DDT belongs) are chemically stable; their survival in soil, water, and living tissue is assured.

(5) Insidious pathways of biological transfer of toxic chemicals are now well known. These channels of potential harm, long suspected by biologists, include delayed toxicity; secondary poisoning; transfer, storage, and concentration along plant and animal food chains; and, at least potentially, mutagenic and carcinogenic effects. Yet many chemically oriented persons continue to disregard the importance of these biological consequences of pesticide use.

(6) Their use is entirely too single-minded in relation to the manifold effects they produce. It is not enough to restrict our concern to their value in crop and personal protection.

(7) The technically supported viewpoints of the conservationist, the resource analyst, the biologist, and the sociologist are too frequently overlooked in pest control recommendations and in governmental programing.

(8) Equally lacking is overt concern for the aesthetic and moral values that must be considered in any application of technology to the satisfaction of human needs and desires.

There are other reasons for complaint and those listed are not to be weighed equally. None denies the clear value of pesticides in safeguarding ourselves and our food supply. All have been debated in the limited circles in which technologists, regulatory officials, agriculturalists, industrial representatives, and governmental policy-makers move. However, the era of closed debate has ended. *Silent Spring* (Carson, 1962) made the debate public property.

The fields, forests, ranges, and waters are living systems capable of change. With wise and cautious manipulation these producing environments will continue to supply our needs and pleasures. Prudently directed ecological disturbances for the special satisfactions of man require no justification. When pesticides are skillfully intercalated into living systems with comprehensive knowledge of their target purposes, the effects they produce, and the values they influence, they too require no justification. Unfortunately, most pesticide practices do not measure up to these standards. We have little enough understanding of the dynamics of biological environments. Yet now we are altering them at a rate that precludes our understanding. Moreover, the appearance of new kinds of pesticides and of new and profound effects from those long in use has outpaced the rate at which their effects can be investigated. The dilemma is a choice of directions. Shall we take the path to achieve what Aldo Leopold (1949) called "... a state of harmony between men and land"—a recognition of the concomitant place of man in living systems? Or shall the path be toward the seeming dominance that chemically maintained, ecological simplification promises?

II · CHEMICALS IN THE
CONTROL OF PESTS

The common and often indiscriminate use of pesticides is perhaps the greatest single contributor to the dilemma. The unfortunate biological results of this pervading use are the preoccupation of this book. But to present these results without some attempt to describe the nature of chemicals causing them, the philosophy and purposes guiding the uses of pesticides, their economic importance, and the manner in which use is regulated is unfair to both subject and reader. Such an attempt is made in the next four chapters. The descriptions are necessarily generalized and purposely subjective. The literature sources cited provide easy entry into the many-sided and difficult problems only hinted at here. The knowledgeable reader may wish to turn immediately to Chapter 6.

2 · AN INTRODUCTION TO
CHEMICAL PESTICIDES

The uses of pesticides in present-day plant and animal protection are far too vast a subject for detailed presentation here. Many hundreds of trade and technical journals devote their attention to the nature, production, marketing, use, hazards, and worth of control chemicals. Thousands of articles on these phases appear each year. It is probably fair to say that the majority of scientists throughout the world are somehow in a broad sense connected with chemical-use programs. Certainly this judgment applies to medical and public-health workers, agriculturalists, and veterinarians. However, technical mastery of the field encompassed in pesticide chemistry and its biological applications is beyond the abilities of most scientists. It is small wonder that scientific advisory and regulative bodies are so fragmented in this area.

Vast as is chemical technology in relation to crop protection and disease control, generalized accounts have been attempted. The best single effort is the book *Insect Control by Chemicals,* by A. W. A. Brown (1951). A book by R. L. Metcalf (1955) is an excellent recent summary, although somewhat narrower in scope. Both authors are entomologists.

The great variety of chemicals and their uses dictates against a single, uniformly applicable classification. I have listed below five general classifications. Each has its value and should be used where appropriate. As will shortly become clear, it is now the custom to consider chemical structure (Table 1) the meaningful center of a classification system, and to make such modifications toward other systems as seems logical. Thus, organophosphorus compounds are chemically identified under their "targets" as insecticides, miticides, or rodenticides. Or, to complicate matters still further, rotenone, for example, may be identified by chemical structure, by its botanical origin, or by its targets, either insects or fish, depending on the manner and placement of applications. The general inference is clear: Physiologically active chemicals rarely produce effects confined to a single species. The manner of use is depended upon to localize effects.

A Classification of Chemicals used in Pest Control

BASED ON PHYSICAL STATE (SOLID, LIQUID, OR GAS)

Solid materials are exemplified by—

(1) Dusts: finely divided particles of the active material, usually combined with an inert bulk filler of equal size and easy dispersibility.

(2) Baits: toxic materials dusted on or soaked into food materials of a target species (e.g., strychnine-dusted root baits for pocket gophers).

(3) Seed dressings: coatings of toxic materials formed from dusts, or from crystals derived from solutions or suspensions in which seeds are soaked. In effect, seed dressings are the same as baits (e.g., fungicidal compounds on seed corn).

(4) Granules: large particles normally containing a mixture of an inactive ingredient with a lesser proportion of the toxic material. Their function is to release toxic materials relatively slowly, permitting residual (long-lived) control.

Liquids are the common "spray" materials, exemplified by—

(1) Toxic substances in true solution (e.g., nicotine sulfate in aqueous solution, or DDT dissolved in acetone).

(2) Toxic substances in suspension: finely divided solid particles in large volumes of water (normally) to facilitate dispersal (e.g., "slurries" of DDT wettable powder in water).

(3) Toxic substances in emulsions: a toxic material usually suspended as fine droplets in water, or dissolved in an appropriate solvent before emulsification in water. Commercial emulsifiers are added to prevent droplets from coalescing. Normally, large volumes of water are used to aid in coverage.

(4) Vapors: finely divided droplets delivered under pressure and held suspended in air, such as in some "mist" spraying or in some kinds of "fumigation."

Gases are reserved for confined spaces (normally limited to stored products).—

(1) Toxic substances applied as gaseous fumigants in storage containers (e.g., carbon disulfide in ground squirrel burrows or in storage bins).

(2) Toxic substances applied as solids but volatilizing quickly (e.g., crystals of paradichlorobenzene in moth control).

(3) Toxic substances, applied in any form, that volatilize to add a fumigation effect to other kinds of contact (e.g., parathion in citrus groves; soil fumigants).

BASED ON TARGET SPECIES

Toxic chemicals can be broadly classified in reference to the plant or animal groups toward which they are most frequently directed or are most successful in control.

Insecticides.—Since most pests are insects, this classification is largest. For convenience, sometimes mollusks, crustaceans, and other arthropods are included as "insects." For example, freshwater crabs in Central Africa are killed with the "insecticide" DDT because they act as intermediate host to the blackfly causing the human affliction onchoceriasis. The crab is the most susceptible link in the vector chain.

Herbicides.—Plant control agents act in several ways, but all have the ultimate effect of killing plants. Selective herbicides affect one major group and not another. For example, 2,4-D acts far more effectively on broad-leaved plants than on narrow-leaved plants. In general, herbicides produce death by direct poisoning or by accelerating growth rates. "Partial death" results when only parts of plants are affected. Examples are the debarking of trees with sodium arsenite and defoliation of cotton to facilitate harvest.

Acaricides and miticides.—Arachnids of the order Acarina include the ticks and mites. Most insecticides do not kill them. The flare-ups in mite numbers in postwar years are directly traceable to the removal of insecticide-susceptible predators of these mites. In recent years a good deal of emphasis has been directed toward mite control.

Fungicides.—The most destructive group of plant parasites is the fungi. Many are beneficial, but many others—rusts, mildews, and molds —are serious plant pests. Fungi are normally controlled by copper-bearing or sulfur-bearing compounds.

Rodenticides.—While these poisons are toxic to all orders of mammals, their predominant use is against rodents and carnivores. The same poisons are also occasionally used to control birds. Traditionally, rodenticides have been chemicals of high toxicity applied to bait materials. Recently, the anticoagulants (e.g., Warfarin) have become the major chemicals to control commensal rodents. The anticoagulants are seldom used away from human dwellings.

All the foregoing groups are classified by pest targets. Sometimes the "target" is defined by crop or commodity type, the commonest example being "cotton insecticides." There are other similar expressions, but such use is better avoided.

BASED ON PURPOSE OF APPLICATION

Direct effects on a pest species are of two kinds: reduction in pest numbers (referred to most often as reductional control) and prevention of access. Reduction in numbers is by far the commoner control purpose; the extreme case of reduction is eradication. Purposeful eradication of well-established species is difficult, expensive, and usually unsuccessful. Preventing access to a crop or dwelling site is normally accomplished by physical means. Examples are trenches around fields to trap white-fringed beetles, and rat guards on ship hawsers. Chemical repellents are, in general, not widely used. In a few special situations (e.g., rodent repellents on pine seed), they yield good results. Cost of treatment is frequently prohibitive, but a more basic reason is the lack of effective repellent chemicals.

Chemical pesticides may be used indirectly to produce either of the results described above. If number reduction is intended, a toxic chemical may be directed at a link in a chain (e.g., mosquito control to reduce transmission of the malarial organism), or it may be applied in a manner that ensures translocation to a part of a plant or animal carrier where it becomes available to the offending pest species. Systemic insecticides are based on the latter idea. Such chemicals are distributed by "circulatory" systems of plants and animals and render the entire organism toxic; hence they have as their targets sucking species or, in a few instances, species living in circulatory fluids.

Hormone herbicides accelerate growth processes and, if the growth rate is extreme, produce death. An important instrument in the success of growth hormones is translocation within the treated plant (see Crafts, 1961).

Chemical attractants are usually a food bait or a sex scent which lures the animal to the means of destruction. Examples are urine scents at coyote trap "sets," and electrocution devices near insect sexual scents. Note here that the most successful attractants are derived from the animals themselves, and depend for their success on detailed knowledge of the animals' habits and physiology (Green *et al.*, 1960).

BASED ON MANNER OF CONTACT OR PHYSIOLOGICAL ACTION

The somewhat superficial distinction between stomach and contact poisons is useful for general purposes. Toxic chemicals that are eaten with food and, upon absorption, cause death are stomach poisons. Producing their effects through the surface of the body qualifies chemicals as contact poisons. Insects absorb contact poisons directly through the outer skin layer or through the respiratory system. Most insecticides are applied to plant surfaces to ensure that the poison will be eaten or walked on. "Tracking powders" in rodent runways accomplish a kind of exposure not dependent on baits.

A time factor is frequently indicated in descriptions of effects. Most commonly, the alternatives are "short-lived" and "residual." No easy distinction can be made between the two. Short-lived or immediate-action chemicals usually retain poisoning capabilities for only a few days; residual chemicals maintain their toxicity for longer periods— usually three to four weeks, but claims are sometimes heard of residual control for one or more years (e.g., control of fire ants for two or three years by granules containing heptachlor). Residual effects are particularly important in control of disease vectors; often-repeated treatments are too costly. In no case do residual chemicals control with equal effectiveness throughout their entire active period.

The pharmacologic actions of a given class of toxic chemicals may be used in description, usually by toxicologists and clinicians. The modes of action of major chemical groups of interest to my subject are summarized below. For further information, see the following references: entomological (R. L. Metcalf, 1955); toxicological (Stewart and Stolman, 1961); wildlife (Rudd and Genelly, 1956).

Organic phosphate compounds.—These poisons act as potent inhibitors of the neural enzyme cholinesterase. Acetylcholine accumulates at the nerve junctions because of the inactivation of cholinesterase and impedes transmission of impulses to glands and muscles. Symptoms of poisoning in mammals include lacrimation, vertigo, muscular weakness, tremors, and dyspnea. In some instances, particularly among the systemic insecticides such as Systox, the technical material does not have as much inhibiting capacity as do its metabolites formed after assimilation into living tissue. There is very little residual accumulation of organic phosphates in tissue although additive effects are possible if exposure is repeated before cholinesterase levels have returned to normal. Aquatic organisms are surprisingly little affected by dosages highly toxic to terrestrial forms.

Chlorinated hydrocarbon compounds.—Unlike the organic phosphates there is no real understanding of the mode of action of the chlorinated hydrocarbon insecticides. Moreover, effects of various members of this class are often so divergent as to suggest quite different modes of action. In general, however, if one uses the best-known member—DDT—as primary reference, the site of physiological action is the central nervous system. Symptoms following single exposure range from increased sensitivity and hyperexcitability with low dosages to characteristic severe tremors, tonic contractions, and convulsions with high dosages. Chronic ingestion results in many histological changes, among which is fatty infiltration of the liver and heart. Residues may be stored, accumulated, and concentrated in fatty tissue to be released slowly when exposure is withdrawn or more rapidly when stored fat is called upon as an energy source. Aquatic organisms are extremely sensitive to most chlorinated hydrocarbon insecticides. Apparently oxygen uptake at the gills is impeded; death comes from suffocation, not—initially at least—from toxic effects on the central nervous system.

Arsenical compounds.—Acute arsenical poisoning is characterized by gastroenteritis and diarrhea leading to violent spasms before death. Kidney and liver injury are common consequences of chronic poisoning. Arsenic compounds with heavy metals will cause symptoms and injury characteristic of both arsenic and heavy metal poisoning. Lead arsenate, for example, may also induce neuritis, kidney degeneration, and vascular disturbances.

Carbamate compounds.—The primary mode of action of the toxic carbamates can, for our purposes, be considered cholinesterase inhibition resulting in symptoms like those described for the organic phosphates.

Rodenticides.—Because of their wide diversity, rodenticides cannot be classified by a single chemical or physiological category.

(1) Strychnine: The spinal cord is affected by acute poisoning producing general stiffening of the body, tonic convulsions, and paralysis. Strychnine is relatively stable in tissue and may result in secondary poisoning to scavengers of the victim.

(2) Sodium fluoroacetate (Compound 1080): The first action of this extremely hazardous chemical is excessive stimulation of the central nervous system. Heavier dosages or delayed actions produce cardiac dysfunction, particularly ventricular fibrillation. Compound 1080 is highly stable in living tissue and regularly allows secondary poisoning.

(3) Zinc phosphide: Depression is a prominent symptom. Poisoned animals are comatose and internally show severe liver damage. Zinc phosphide is not as stable in tissue as strychnine and Compound 1080. The odor of phosphine gas can be detected arising from the intestinal contents of poisoned animals.

(4) Thallium: The primary action of thallium is on the central nervous system. Initial symptoms of acute poisoning are depression, lack of muscular coordination, and respiratory distress. Chronic ingestion yields severe effects on the liver, kidneys, and alimentary canal. In spite of its high toxicity, thallium kills rather slowly. Most fatalities occur within three days of ingestion of a lethal dose. Thallium is stable in tissues and has regularly produced secondary poisoning.

(5) Warfarin: This relatively safe chemical depends on repeated ingestion for its action. It is an anticoagulant, its action being to lower the prothrombin (a blood-clotting factor) content of the blood. All affected animals show severe internal hemorrhaging.

BASED ON CHEMICAL NATURE

The most useful system of pesticide classification defines detailed chemical character within a loose framework of control purpose. Thus, under "Insecticides" we identify inorganic and organic chemicals. In-

TABLE 1

A condensed classification of major pesticides

Chemical group or action	Examples
Insecticides and acaricides	
Inorganic	
Arsenicals	Lead arsenate
Copper-bearing	Copper sulfate
Sodium fluoaluminate	Cryolite
Organic, naturally occurring	
Nicotine alkaloids	Nicotine sulfate
Pyrethroids (also synthetic)	Pyrethrum
Rotenoids	Rotenone, ryania
Organic, synthetic	
Chlorinated hydrocarbon compounds	Aldrin, aramite, benzene hexachloride, chlordane, DDT, endrin, heptachlor, methoxychlor, toxaphene
"Dinitro" compounds	DNOC
Organic phosphorus compounds	Chlorthion, DDVP, diazinon, malathion, parathion, phosdrin, schradan, TEPP
Carbamates	Sevin
Fungicides	
Mercurials	Mercuric chloride, "Ceresan"
Quinones	Phygon
Dithiocarbamates	Nabam, ziram
Others	Captan, thiram
Herbicides	
Contact toxicity	Sodium arsenite, oils, "dinitro" compounds
Translocated (particularly hormone types)	2.4-D, 2.4.5-T, dalapon, many others
Soil sterilants	Borates, chlorates, others
Soil fumigants	Methyl bromide, vapam
Rodenticides (mammal poisons)	
Anticoagulants	Warfarin, pival
Immediate action	Strychnine, sodium fluoroacetate ("1080"), phosphorus, ANTU, thallium, endrin
Other vertebrate targets	
Birds	Strychnine, TEPP
Fishes	Rotenone, toxaphene

organic chemicals may be further divided into compounds containing arsenic, copper, mercury, and so on. Organic chemicals may be handled in similar fashion. The final classification reflects the major use of the chemical, but does not avoid some overlap among the various categories of control purpose. For example, endrin is a synthetic chlorinated hydrocarbon useful as both insecticide and rodenticide. Table 1 presents the major pesticides according to a simplified "mixed" classification.

Assessing the Hazards of Chemicals

Many thousands of chemicals have been tested as pesticides. Any accepted chemical must serve a recognizable control purpose, must have no uncontrollable side-effects on human beings, must be usable in present productions systems, and must be available at reasonable cost. Not many chemicals satisfy these requirements. The manner of achieving the control purpose is not critical; there are many modes of action. Unusually high inherent toxicity is not of itself a deterrent to pesticide acceptance. More important considerations are the degree of success in control and the magnitude of side-effects. Some highly toxic chemicals are used with greater safety than some of lesser toxicity.

Toxicity and hazard of chemicals are not synonymous. Pesticides are by definition toxic agents. A pesticide is usually capable of harming forms of life other than the target animal. Very few application techniques can be considered thoroughly safe as far as potential hazards under all conditions of use are concerned, and therefore most pesticides should be regarded as hazardous. The same applies to medicinal chemicals prescribed by physicians. The potential hazard of any chemical varies directly with the character of its toxicity and with the level of dosage: The hazard may be slight, as with the hormone herbicides, or it may be great, as with most rodenticides. In practice, however, this interpretation is only partially correct. Formulation and use recommendations alter both toxicity and exposure. The manner of application and the area in which a pesticide is applied are more important considerations in assessing actual hazard than is inherent toxicity alone. Generalizations regarding hazard based solely on toxicity are better avoided. Hazard judgments must consider inherent toxicity in the context of local use and environment. Only then can actual hazard be determined.

For practical purposes, actual hazards cannot be assessed until a chemical is in use. The sequence followed in determining the efficacy and hazard of a pesticide is about as follows:

(1) Screening of chemicals yields promising "candidate materials."

(2) Immediate toxicity of purified material is gauged in laboratory animals by single administrations at different dosage levels through oral, dermal, and respiratory routes. The individual result is the acute toxicity. Because responses to toxicity vary greatly, individual toxicity values are judged in relation to a large test population. Toxic symptoms and sublethal effects are also noted at this stage of testing. Death is the chief criterion. The effectiveness of the chemical at a given dose is gauged by the number of deaths in a test population. The most common standard for comparison is the LD_{50} (lethal dose for 50 per cent), the amount of chemical resulting in the death of half the test population. These values for many common pesticides discussed in this book are presented in Table 2. Note that these values are approximations at best. They do not provide comparisons of differential effects deriving from age, sex, and species, nor can they be used to interpret many of the subtler but important effects exemplified later. There are also other useful percentage figures, notably the LD_{100}, the minimum amount of chemical that kills all the test animals.

(3) Subsequently the effects of multiple doses and long-lasting effects of single doses are observed on large populations, usually over a period of a few days to a month. However, these observations may be regarded only as short-term toxicity studies. True chronic toxicity must be observed over considerably longer periods—up to perhaps two years. With the current emphasis on chemical residues, chronic-toxicity studies normally center on continuing intake of small amounts of chemical, a single dose of which would produce no effect (see Table 2).

(4) Before chronic studies have been completed, small plot studies are begun at the research stations of the manufacturer or at those of governmental agencies or universities. The agricultural experiment stations are important cooperators at this phase. At this stage the candidate chemical is prepared in formulations which are appropriate for pest control under field conditions.

(5) Continued "promise" being assumed, the chemical is applied in different formulations under actual growing conditions on croplands. Many farmers cooperate in this phase. There is concerted effort at this time to apply the chemical on as many crop types and in as many regions as possible.

TABLE 2

Acute and chronic oral toxicities of some common toxic pesticides, based on experimental studies with rats. Chemicals are arranged in decreasing order of acute toxicity within major headings. Data from A. W. A. Brown (1951), Martin (1957a), Metcalf (1955), Rudd and Genelly (1956), and Spector (1956). '

Pesticide	Acute LD_{50} (mg./kg.)	Hazardous chronic level (p.p.m.)
Arsenicals		
Sodium arsenite	10	—
Arsenic trioxide	138	15
Lead arsenate	825	<4
Botanicals		
Rotenone	132	30
Ryania	1,200	10,000
Pyrethrins	1,500	50
Organic phosphates		
TEPP	2	—
Thimet	4	—
Parathion	6	25
Phosdrin	7	30
Schradan	10	1
Chlorthion	880	75
Malathion	1,375	5,000
Organic carbamates		
Sevin	500	>750
Chlorinated hydrocarbons		
Endrin	10	5
Heptachlor	60	40
Aldrin	67	25
Toxaphene	69	1,200
Dieldrin	87	10–50
BHC (Lindane)	125	400
DDT	250	100
Chlordane	335	150
Methoxychlor	7,000	5,000
Dinitrophenols		
DNOC	26	100
Rodenticides		
Sodium fluoroacetate	3	—
Strychnine alkaloid	16	200
Thallium	17	50
Zinc phosphide	40	100
Warfarin	>200	6
Petroleum distillates		
Xylene	—	200 (air)

(6) Collected information is presented to regulatory agencies for registration. Information includes toxicity and field-trial data and any recommendations for specific uses. In most cases, the testing time is insufficient for complete chronic-toxicity data, and use recommendations are meager.

In general, the foregoing six points cover the sequence of development and first use of a pesticide. At the completion of this sequence, the pesticide is put on the market if there are no apparent contraindications arising from the testing. A close watch is kept initially on the manner and quantity of consumer use, while long-term chronic-toxicity tests continue and tests are made on other species of crop and pest. In short, to the point of marketing, hazard is assessed largely by short-term toxicity trials, and efficacy in pest control is judged primarily on limited field trials. Manufacturers, growers, and public-health and regulatory officials depend on further consumer use to refine the assessment of actual hazard. Undesirable side-effects in nature are rarely evaluated and can only be inferred from accumulated data. Dependence on use to assess full hazard is the rule, not the exception. This state is looked upon as reasonable and desirable by most individuals involved in developing and testing.

3 · DOMINANCE OR CONCOMITANCE?
LOCAL, MASS, AND INTEGRATED CONTROL

The target being a pest species, either plant or animal, there are two courses toward effective control. One route to its control is a chemical blanketing which discounts the place of natural regulating forces in the biotic community. The alternate route places its primary dependence upon natural regulating forces and sparingly intercalates toxic chemicals as supplements when necessary. Most local control and what has come to be known as "mass" control follow the first route; "integrated" control, the second. One route rests on man's absolute dominance of the natural biota through chemicals, satisfying only a single purpose; the other, on concomitance with the biota, assisted by chemicals, satisfying many values. This chapter dwells at length on these alternatives for they are the basic choices facing present pest control practice.

Chemical Blanketing—Local and Mass Control

Both local and mass control depend on the same manner of chemical distribution. Yet there are some important differences in both the purposes and the effects of application. Traditional pest control is local, applied to single crops or environments, dominated by private interests, and scarcely capable of blanketing large areas. Mass control is very nearly the opposite of each of these points. At its simplest, mass control means large-scale pest control, but the size of the treated area is not enough alone to differentiate local or traditional pest control from mass control. Purposes, sponsors, methods, hazards, and attitudes also differ. The following parallels will identify the differences between local and mass treatments.

Local	*Mass*
1. Target of control usually a single pest species.	1. Target also a single pest species, but more widely damaging.
2. Normally, an established pest species responding to usual cultural practices.	2. Normally, a recently established, rapidly dispersing species favored by both cultural practice and climatic sequences.
3. Treatment confined to restricted environment, usually a single crop type.	3. Treatment covers many kinds of environments, although one type may predominate.
4. Control result usually temporary suppression of numbers, necessitating repeated treatment.	4. Result usually suppression of numbers for periods longer than a year.
5. Purpose very rarely quarantine or eradication.	5. Purpose frequently quarantine or containment, often with eradication as a stated goal.
6. Nonselective chemicals applied as a rule; but selectivity of target often achievable.	6. Similar chemicals used; but selectivity rarely possible.
7. Application sometimes by air, but ground treatment usual; normally, treatment not coordinated with that on neighboring areas.	7. Application normally by air, often using the larger aircraft not practical in local crop treatment. Where a large-scale control campaign can not rely on aircraft (e.g., in Dutch elm disease control), coordinated programing achieves the "mass" effect.
8. Treatment initiated and paid for by landowner; or if, as in mosquito control, responsibility not assumed by landowner, operation initiated and paid for by local government.	8. All programs conducted by governmental groups, normally state and federal; public funds meet at least two-thirds of cost.

Local	*Mass*
9. Exposures to toxic chemicals acute and temporary.	9. Exposures may be acute, but are more apt to be chronic and longer lasting.
10. Faunal instability is inherent in local treatment areas; chemical application merely accentuates it.	10. Faunal instability derives chiefly from chemical application.
11. Recovery of desirable species more likely in limited, local treatment.	11. Recovery of desirable species slower, because of reduction over larger areas; sometimes compounded by chronic effect.
12. Elimination of desirable species not likely directly from chemical treatment.	12. Virtual elimination of desirable species as result of mass control. After all, the purpose of mass control is often eradication— induced extinction. The measures to accomplish this in one species do the same to other species.
13. Local control is generally conceded to be necessary to plant and animal protection and to public health.	13. The necessity for mass control is not so conceded.

Differences between the two types of control are not absolute. There is in fact a transitional level, where differences are hard to define. Transitional stages come about for these reasons:

(1) Private control programs, through either custom or active coordination, result in synchronous practices over contiguous areas. One example is trapping, shooting, and poisoning of predatory mammals on private rangelands.

(2) Occasional pest outbreaks are countered by single practices throughout a large area. An example is the spraying of dieldrin over 130,000 acres of California rice land in 1953 to suppress an outbreak of rice leaf miner.

(3) Regular and repeated recurrences of pest outbreaks in single crop species met by uniform methods of control result inadvertently in wide-scale synchronous treatment. Repeated parathion treatments in apple or citrus orchards will illustrate.

(4) Regular dependence on common advisory sources such as university extension services results in standardized pest control proce-

dures uniformly affecting large areas in which similar crop types occur. Points 3 and 4 are obviously related—the latter being the source of recommendation, the former the recommendation in practice.

Some common practices therefore do produce a kind of mass control similar in effect to planned mass control. In few instances, however, are area-wide effects considered a goal of treatment or, for that matter, recognized for what they are. Intent is an important component of planned mass control; the purposes of mass control and local control are not the same. The hazards of mass control (however control is intended) include those in local control, although frequently intensified, along with others peculiar to mass control. Examples of the hazards of mass control are resistance, residue contamination, delayed and sublethal effects, and threatened extinction of native species. These and other problems are discussed in detail later. At this point I wish to indicate some common control programs that are based on mass-treatment methods. The first five of these are described later as type examples of the scope, methods, and hazards of mass-control operation. These five campaigns alone have in recent years annually covered between four and six million acres in the United States and Canada. The total annual world-wide coverage of the campaigns listed below (and the list is incomplete) probably exceeds 200 million acres.

> Spruce budworm (U.S., Canada)
> Gypsy moth (northeastern U.S.) and
>   several other forest insects as well
> Japanese beetle (mid-Atlantic, central U.S.)
> Dutch elm disease (northeastern, central U.S.)
> Imported fire ant (southeastern U.S.)
> Grasshopper (western U.S.)
> Mediterranean fruit fly (southeastern U.S.)
> Mormon cricket (western U.S.)
> Mosquitoes (many countries)
> Beet leafhopper (California)
> Tsetse fly (Africa)
> Deer and opossum (New Zealand)
> Forest rodents (U.S.)
> Wild dogs—coyotes, dingoes (U.S., Australia)

SPRUCE BUDWORM

This tortricid moth and some closely related forms occur in northern coniferous forests of the Pacific slope and in the northeastern United States and bordering Canadian provinces. In recent years, its depredations have been most severe in Maine, Quebec, and New Brunswick. The larvae are defoliators; when present in large numbers, they are capable of killing trees. Partial destruction of foliage, moreover, slows the growth of timber and in some instances makes trees more vulnerable to additional insect attack. In contrast to most major pests, the spruce budworm is native to this continent. Damage is not severe in the southern reaches of its distribution, where forests include deciduous trees and where natural enemies apparently limit population numbers effectively. Where outbreak levels occur to the north, spruce and balsam predominate. The species seems to be curiously out of balance with natural controlling factors for sequences of several years. Sometimes the build-up and maintenance of high populations, continuous for 10 to 15 years, appear to be associated with favorable long-term climatic sequences. The species is, in fact, spreading its range, and is doing so without special assistance from man's timbering activities. Apparently the dispersal and high population densities of the budworm reflect the biological responses of a species not yet in equilibrium with its environment nor yet having reached its ultimate distribution.

Spruce budworm damage has been recognized as severe for over fifty years. Studies (notably by F. C. Craighead and S. A. Graham) beginning in the 1920's set the pattern for increasingly intensive investigations, which have continued to the present. For several years the Forest Biology Section of Science Service in Canada and the United States Forest Service have maintained well-supported field and laboratory investigations on the budworm. Particularly important research centers are the forestry laboratories at Yale University and the University of New Brunswick. It was not until 1944 that trials with DDT revealed any genuine promise of area-wide control of the budworm. Very intensive trials quickly followed, and within three years aerial applications of DDT over forests became the standard method of control of budworm and several other forest insects. Up to five million acres of conifer forest have been sprayed in single years with DDT in attempts to suppress the budworm. Ordinarily the treatment is made on blocks of thousands of acres with DDT in oil at rates of 0.5 or 1 pound per acre. Much of the control operation takes place on public or crown lands, and is planned and conducted by public agencies. Public funds meet much of the cost.

The experience of the past fifteen years confirms that direct loss of terrestrial animals is unlikely if recommended procedures and dose rates are observed. Aquatic animals, however, are particularly sensitive to DDT, and catastrophic reductions of fish and aquatic insects often follow DDT contamination of streams. Less direct effects of DDT spraying of this kind do occur on both land and water. Some of these I describe later (see Chapter 20, for example), but the full extent and significance of these effects (one must fairly report) have yet to be assessed.

Spruce budworm operations with DDT have probably covered more continuous acres than has any other spraying scheme, possibly 20 million to date. From the beginning the operations have been questioned by conservationists on the basis of actual or potential hazard. More recently, they have been criticized by forest entomologists and managers, who have concluded that inordinate dependence on chemical control does not permanently solve the control problem, and that such dependence in fact inhibits efforts to install good silvicultural and biological practices that would be permanent controls. Whatever the argument, aerial spraying for budworm will probably remain until an effective biological antagonist can be found and distributed. I suspect that many years and millions of treated acres separate the present and the time when the budworm will be permanently suppressed by nonchemical means.

GYPSY MOTH

Control of the gypsy moth, like the spruce budworm, depends on relatively light applications of DDT in oil (1 pound per acre) aerially dispersed over wide areas. As with the budworm, gypsy moth larvae defoliate trees and, when present in outbreak numbers, are quite capable of killing trees. Public agencies normally conduct control operations against the gypsy moth. In fact, for seventy years state and federal agencies have recognized the seriousness of the gypsy moth as a pest and have organized publicly financed campaigns against it. The gypsy moth may thereby claim the questionable honor of being the first forest insect against which mass-control operations were conducted. These points end the similarities to the spruce budworm and its control. The differences between control procedures used against the two species have given substance to one of the bitterest controversies yet to arise in pesticide use. Before I elaborate on the reasons for this strife, a few facts are needed about the history of gypsy moth control itself (see Worrell, 1960, for a particularly thorough review).

Gypsy moths in this country derive from a few individuals that in 1869 escaped from a colony near Medford, Massachusetts, that had been brought to the United States from France by a French scientist. Within twenty years the population had grown large enough to cause serious damage to hardwood trees, and in 1890 the Massachusetts legislature enacted a law aimed at eradicating it. Aggressive spraying campaigns followed, which were apparently effective in reducing numbers. But adverse weather over a period of years contributed to low numbers. By 1899 the moth was seemingly producing no further damage, and support for control was withdrawn. However, the insect population quickly built up again, and by 1905 the gypsy moth was more widely established than before. In that year the Massachusetts legislature renewed support, and establishment of the moth in other New England states precipitated the intervention of the federal government. Within a few years the insect had spread into New Jersey, New York, and Pennsylvania, but in these states most pockets of infestation were eliminated by local eradication programs. Attempts to import natural enemies led to successful establishment of a few but did not effectively curb either population numbers or the westward dispersal of the insect. Two standard control procedures were followed in the 1920's and 1930's. One was chemical treatment of outlying pockets of population at the strict quarantine or barrier zone surrounding the main population center. The purpose was containment until effective biological control agents could be found. Outbreaks of moths within the containment area were, of course, treated with insecticides (normally lead arsenate). This policy continued for some twenty years. Unfortunately, the strong hurricane that struck New England in 1938 carried large numbers of insects into southern New York State and into Pennsylvania, where earlier establishments had been removed. Populations larger than previously known were now securely established beyond the barrier zone.

The appearance of DDT changed the outlook from one of containment to one of eradication. The hazards of DDT were not well known at the time, however, and efforts to learn of them led to several years of the most intensive studies ever conducted on the side effects of a chemical in field use. Joint studies by the Department of Agriculture and the Department of the Interior continued from 1944 to 1949. The effort was commendable and the results were comprehensive; like ventures have not been made since. The gypsy moth reached Michigan during this time, and the earlier fears of entomologists were confirmed. There seemed to be no containing it. Eradication was the only answer.

Not until the middle 1950's were the study, support, and machinery of control (with eradication as its purpose) sufficiently integrated to initiate mass control over large areas. While many complaints had been lodged against small-scale campaigns conducted over the years, most of the complaints were considered minor by control officials, and plans to increase control coverage went ahead without delay. The plan for 1957 was to spray most of a 2.5-million-acre area.

Objections were formalized during 1957. In May of that year a group of Long Island residents, headed by the eminent ornithologist Dr. Robert Cushman Murphy, sought a temporary injunction against the spraying of their properties. The injunction was denied and the 1957 campaign went forward. But spraying was not confined to woodlands. Unlike previous federal mass-spraying schemes, the program was in large part conducted in densely settled areas and also contacted pastures and forage crops. It received a great deal of unfavorable publicity. The outlying infestations apparently were eliminated in 1957.

Meanwhile the Long Island group had appealed its case, asking for a permanent injunction against the spraying on its properties. With court action pending and the 1957 campaign an apparent success, control officials decided to hold eradication programs in abeyance during 1958.

Testimony in the "gypsy moth" or "DDT" trial was offered before Judge Bruchhausen in February and March of 1958. Very soon it became apparent that the entire issue of aerial spraying with DDT was on trial, not simply the violation of individual property rights. Some 1400 pages of testimony by 50 witnesses reviewed the relevant questions of hazard to human health and wildlife, of control philosophies and successes, and of the general advantage served. In June, Judge Bruchhausen dismissed the request for injunction on the basis that charges against the program had not been substantiated.

The pesticide industry and the control agencies of government had "won." What they won, however, was a technical—only a legal—decision. The publicity attending the trial alerted many thousands of people to the potentialities for harm that such control campaigns offer. This fact was recognized by control officials. One practical manifestation was the decision of the Plant Pest Control Division not to continue its eradication attempts against the gypsy moth in the same fashion.

The gypsy moth has fortunately "cooperated" to the extent that outbreak levels have not reappeared. What will happen when an outbreak reappears is difficult to predict. It is sure, however, that mass treatments will in the future be subject to closer scrutiny by control agencies themselves; the sensitizing of large segments of the public has ensured that. Thus, the final decision on the "DDT trial" may yet be rendered.

JAPANESE BEETLE

The first American appearance of the Japanese beetle was in New Jersey in 1916. In its native Japan it is not considered a pest, although it has become one when transported elsewhere throughout the world. The beetle has flourished in this country, spreading in forty years throughout most of the states east of the Mississippi River. It is most favored where summer rainfall is moderate and winter temperatures do not stay below freezing for long. The insect is capable of strong flight, and its gradual westward spread has been largely the result of its natural ability to disperse, though commercial transport also favors it. It has been recorded in isolated colonies in many places (mainly airports) throughout the United States. A good deal of furor, for example, followed the recent discovery of several individuals near Sacramento, California, a full 2000 miles from the nearest center of high population.

The most accessible link for control in the life cycle is the grub, or larva, of the beetle in soil and turf. Ordinarily, therefore, insecticides at high concentrations have been used to contaminate the top few inches of soil. General soil contamination demands higher amounts of insecticides than does foliage treatment. Over the years of chemical treatment of the Japanese beetle, amounts of insecticides have always been high and hazardous. Per acre figures have, in a succession of recommended procedures, ranged from 400 pounds of lead arsenate, to 25 pounds of DDT, to 2 pounds of aldrin. In spite of the apparent differentials in dose rates, the effectiveness of these amounts is much the same. The hazard seems to have mounted. Now the custom is to apply dieldrin, aldrin, or heptachlor by air at rates of 2 or 3 pounds per acre. At these levels, all are hazardous to warm-blooded animals; moreover, all form persistent toxic residues. Dieldrin is most stable; aldrin converts to dieldrin; and heptachlor forms a stable, more toxic epoxide. Current Japanese beetle control methods, therefore, are inherently hazardous in both the short and the long term.

Recent control campaigns are designed to stop the westward spread of the beetle rather than to eradicate it. In the areas of longest occupancy in this country, Japanese beetle populations seem to have stabilized. Natural enemies have been introduced; a few have been fairly effective in limiting numbers. The best known of these is a bacillus (*B. popilliae*) that causes "milky spore" disease. Under certain circumstances this disease becomes established and keeps beetle numbers well in check. More often, however, effective control requires that the spores be released repeatedly at the correct time. The manner of use is much as with an insecticide. Thus manipulated, milky spore disease effectively maintains low beetle numbers. In fact, the disease was the first to be termed a "living insecticide."

Needless to say, biological control of this kind is not compatible with an eradication philosophy. Eradication demands chemical treatments. Recent efforts to eradicate populations on the western fringe of the Japanese beetle's range have depended on high doses of the chlorinated hydrocarbon insecticides mentioned. Up to 1960, six states at the current western line of distribution (Michigan, Indiana, Kentucky, Illinois, Iowa, and Missouri) had treated 96,000 acres in cooperation with the Plant Pest Control Division of the United States Department of Agriculture. The kinds and amounts of insecticides, the manner of treatment, and the character of the landscape have ensured wildlife loss. Hazards of other sorts may also be expected under these conditions, but most of them are reduced to a minimum through careful observation of proper procedures.

Wildlife losses have been severe on occasion. For example, ground-feeding birds were virtually eliminated from treated areas near Sheldon, Illinois (Scott *et al.*, 1959). So too were muskrats, rabbits, and ground squirrels. No one knows the full extent of loss, nor has anyone studied the more subtle hazards that are likely following mass application of stable chemicals. Not knowing the full impact of such chemical treatments is as much the present cause of concern as is the observed heavy wildlife mortality. There lies the crux of the controversy over current Japanese beetle control campaigns. Comparable treatments elsewhere suggest that wildlife effects may be much greater than immediate mortality alone suggests.

There are other complaints too. One involves the eradication purpose itself. Biological methods can be, and have been, used successfully. Proponents of this method of control essentially argue for "living with" the insect, just as we have been doing for 45 years. They argue that it is too late to attempt eradication; the species is now too well established. Wildlife loss, other hazards, and the costs of successful eradication do not justify an all-out eradication campaign with present methods of treatment.

There are also complaints against the manner of administration of control campaigns. Probably the most serious is the failure to inform the public properly. Also, many areas containing only a few beetles were nonetheless treated as if beetles occurred everywhere in high numbers. This indiscriminate feature of application sustains biological objections to mass-control methods.

As currently practiced, the Japanese beetle campaign appears doomed to fail. It is neither subtle nor imaginative; nor is it (and this is most important) effective. Success in local suppression perhaps retards dispersal, but in no way does eradication appear possible. Two alternatives are open: One is a change of attitude—away from the eradication purpose; the other is a change of method—to specific means of control. A specific chemical means of eliminating Japanese beetles entirely without harm to other animals has yet to be found. Containing its numbers with biological control agents seems the only reasonable solution. I am confident that the public will not tolerate an indefinite continuation of the present chemical approach to Japanese beetle control.

DUTCH ELM DISEASE

The fungus that causes Dutch elm disease has been in this country for some thirty years. In that time it has caused the death of many thousands of urban shade trees. Curiously, the fungus, itself imported from Europe, depends for its transmission on a bark beetle accidentally introduced some years before the fungus made its way to the United States. Both fungus and beetle can live separately, but together they produce "ideal" conditions for their own welfare. A closely related, native beetle also can carry the fungus, but is considered not so dangerous as the introduced form.

Elms are, of course, valuable shade trees. Exactly how valuable in dollars is hard to say. Perhaps three to five hundred dollars is a proper range for an established elm in a good location. But the value of elms transcends economics and involves the intangible values of providing shade, ornamentation, tranquility, and a wildlife haven. Evaluation of loss of such a noncommercial commodity must necessarily be subjective when cost of treatment is weighed against gain from control.

The spread of Dutch elm disease can be slowed by removing all dead wood from affected trees. Beetle galleries are found only in the dead and dying portions of the tree. However, sanitation of this kind is expensive and, in the experience of urban controlling agencies, not likely to be followed consistently by homeowners. DDT is very effective in killing bark beetles and in keeping them out of trees for some months following heavy application. Eliminating the beetle carrier stops the dispersal of fungus spores, and the disease thereby becomes inconsequential.

Since the advent of DDT as the major chemical in the control of Dutch elm disease, the equivalent of over two million acres has been sprayed. Doses per tree range from a minimum of 2.5 pounds of DDT to over 17 pounds where elms are large and numerous (George, 1959). Normally these amounts are dispersed in large volumes of water by high-pressure spraying machines.

In contrast to other large-scale types of treatment, Dutch elm disease control with DDT does not depend on dispersal of insecticide by airplanes. DDT spray is therefore reasonably well confined to the area where elm trees occur. Spraying has probably thus far slowed the spread of the disease. It has nonetheless made its way throughout New England and adjoining Canada, the Middle Atlantic states, and into the North Central states. It has reached as far west as Kansas and probably will continue its spread. The aggregate area under treatment —entirely urban—is therefore very large. The uniformity of recommended procedures and the intensity of their application throughout this large area qualify this manner of treatment as true mass control.

The quantities of DDT necessary to kill bark beetles are more than sufficient to kill birds in or about sprayed trees. Moreover, the stability of DDT ensures its survival for months and even years in the soil under treated trees. This stability has other distressing biological consequences which are later described in detail (see particularly, Chapter 20). I need only say here that bird populations after spraying are often depleted from acute exposure to DDT, and that longer-range effects further reduce bird numbers. Losses are most marked where the density of elms is high. There is no doubt in my mind that millions of birds have been killed by DDT sprays on elm trees. Nor can there be any doubt that the character of urban bird populations must change, at least where elms are common, with repeated heavy spraying for Dutch elm disease control.

Control of Dutch elm disease with DDT has always provoked controversy between bird lovers and control officials. Recently the controversy has been strongest in Michigan, Wisconsin, and Illinois (Hickey, 1961). The problem is paradoxical. Elm trees are appealing and valuable; so also are birds. In saving the elm by current control measures, we preserve a refuge for the bird but we kill the bird. The only logical position is therefore to seek methods that safeguard both trees and birds. Fortunately there exists a chemical near-relative of DDT—methoxychlor—that is about as effective as DDT in killing beetles but poses no real hazard to birds (Wootten, 1962). Unfortunately, methoxychlor costs three times as much as DDT. Until the last year or two this cost differential (one dollar per tree with DDT to three dollars per tree with methoxychlor) was enough to exclude methoxychlor from control programing. However, several communities have elected the seemingly more expensive course on the basis that greater value accrues in the long run. Doubtless other communities will make the same decision and perhaps ultimately DDT will not be recommended for Dutch elm disease control. Other chemicals are under study as control agents, but none yet satisfies the requirements of both bird and tree protection. Few will accept the premise implicit in using DDT on elms: that most birds must be sacrificed to protect trees.

IMPORTED FIRE ANT

Seven species of fire ants (*Solenopsis*) are distributed throughout warm temperate regions of the Americas. Three are native to the United States. Shortly after World War I a dark form of *S. saevissima* from Argentina became established in Mobile, Alabama, and slowly spread throughout the surrounding region. In the 1930's a smaller reddish form of *saevissima* appeared in the same area, apparently representing a second major introduction. The latter form aggressively made its way beyond the state's borders by 1940, swamping out by interbreeding and internecine warfare the established black phase (Wilson and Brown, 1958). At present North American populations are chiefly the red phase, tending to replace native forms as well as the dark phase.

Undisturbed ants build small mounds of hardened earth which become more numerous with increased time since establishment. Older mounds may number 15 to 25 per acre and be two or three feet high.

Needless to say, the presence of mounds interferes with farm operations. Moreover, fire ant stings are painful and two or three deaths have been recorded from their stings. The ant is clearly a major nuisance. However, its role as major crop pest is disputed. Studies in 1948 and 1949 announced hazard to livestock and certain crops, but later appraisals do not confirm predictions from those studies. All research work on the species was conducted during 1948–49 (W. L. Brown, 1961).

Meanwhile in the years from 1940 to 1956 the fire ant had made its way throughout eleven states in the Southeast. Because of the early predictions, although lacking proper study of both ant and control methods, the U.S. Department of Agriculture in 1957 requested congressional approval for control of the fire ant. The request was granted and a 2.4-million-dollar annual allocation approved with the stipulation that local matching funds be made available. Most states and many local agencies and individuals did respond with matching funds, but very often control operations went forward without them. A massive operation was set up with great speed. The first spraying using 2 pounds of dieldrin or heptachlor per acre began in November, 1957. Heptachlor at 1 to 1.25 pounds per acre was substituted in 1958, and this regime was later amended to two applications of 0.25 pound each spaced three or four months apart. Over two and one-half million acres of land have been aerially treated with heptachlor or dieldrin in the dosages described.

Not until the operation was under way were wildlife and health authorities notified. Immediate opposition to the program resulted— opposition which was to grow to national controversy. Fish, wildlife, livestock, and poultry suffered losses, the destruction of wildlife bordering on catastrophic. An initial appraisal (George, 1958), setting forth an outline of observed and probable adverse effects, was later abundantly amplified with data on excessive mortality (DeWitt and George, 1960; Southeastern Assoc. Game and Fish Comm., 1958, among many others). An Agricultural Research Service publication (1958) on the "facts" of the eradication program was countered easily by different and more impressive "facts" (e.g., Arant *et al.*, 1958). The insect was not considered more than a nuisance in any of the southern states; it did not destroy crops, wildlife, and livestock. However, the chemicals in use did eliminate vertebrates from some areas, recovery of some populations not being certain yet. The spraying program did create residue problems; it did contribute to insect outbreaks themselves requiring control (Long *et al.*, 1958); and it has cost at least 15 million dollars. It did not eradicate the fire ant; the ants have in fact "re-infested" most treated areas. The Plant Pest Control Division of the U.S. Department of Agriculture had clearly made a massive mistake; the operation was a failure from its inception.

On the basis of the many doubts engendered, the Alabama legislature withdrew matching funds in 1959, followed by the Florida legislature in 1960. The "all-out attack" was abandoned; local control is present practice. New and effective control methods have been developed under the stimulation of controversy, but they did not come from the control program's sponsor. For example, Hays and Arant (1960) suggested a bait material that seems effective. In 1962 a new chemical (Mirex) was given extensive field trials. This chemical applied on bait seems to control fire ant numbers without obvious damage to game birds (Baker, 1963; George, 1963). Nonetheless, several hundred thousand acres were treated with heptachlor in amounts to two pounds per acre in Louisiana alone in 1962. Full substitution of Mirex for chlorinated hydrocarbon insecticides was scheduled to begin in 1963 (Keith, personal communication).

Unlike the successful eradication programs against the Mediterranean fruit fly and the screw-worm, the fire ant program lacked imagination, depth of prior knowledge, and the promise of success that both inspire. Lacking from the beginning was research information adequate to guide a control program of such scope (W. L. Brown, 1961). Equally lacking was a truly cooperative approach among agencies concerned with the operation. I believe that the failure of this program marks the end of eradication attempts by "blitz" methods.

The Integrated Control Concept

Natural forces restrict the numbers of plants and animals, and these limiting forces affect pests in the same way as any other species. Natural controlling forces do not always keep numbers low enough for man's convenience. Man-induced disturbances, in fact, ensure that pest numbers will on occasion be high. Because of these disturbances, natural means of control cannot in most crop ecosystems be depended on at all times to suppress numbers of pests. Therefore, natural antagonisms may be consciously implemented through biological control. At the present time, chemical and biological control methods are regarded as the chief means of holding pest numbers in check. These two approaches to control are frequently considered alternatives, one method in practice being antagonistic to the other. This supposition of conflict is not realistic. In the first place, it overlooks the fact that natural controls must inevitably be depended on to a large degree in any crop situation, whatever chemical practices are followed. Secondly, with adequate knowledge and conscious management, one method may well supplement the other. The conscious harmonizing of chemical and natural control methods against pests is what is meant by integrated control (general references, Stern *et al.*, 1959; van den Bosch and Stern, 1962).

Thus, integrated control is distinguished from other methods of control by three features existing together in combination: The basic dependence is on natural processes in pest control; chemical and natural methods of control are combined against single pests; and the two kinds of control are integrated in a sequence best suited to the crop ecosystem. With these as operating premises, it is clear that any chemical used in an integrated control program must disrupt natural equilibria as little as possible and should survive no longer than necessary. The general purpose of chemical agents in such a program is to reduce pest populations to levels at which antagonistic organisms take over the function of control. Needless to say, the chemicals must do little or no harm to these competing organisms.

Integrated control is not a new idea. Foresters, particularly in Europe, have used the concept for many years. In fact it forms the basis of most control practices in present European forestry. In Europe also, the concept extends to agriculture, where a thread of minimum biological disturbance runs through most crop-protection practices. The long-term studies of Pickett and his colleagues (e.g., Pickett, 1961) in the orchards of eastern Canada are based on harmonizing chemical and biotic controls for most effective insect control. They are integrated controls, although not so termed. The recent popularity of integrated control among economic entomologists in the United States does not mean that the concept is either new or previously misunderstood. Rather, present failures of chemical control have engendered new enthusiasm for integrated control.

Integrated control is biological common sense. The changes in faunal balance that are inherent in crop ecosystems certainly cannot always be contained satisfactorily by biological means. To combat pest problems economic entomologists have always sought methods that were sure and widely applicable. Chemical controls have repeatedly shown their value in limited control situations. When the high insecticidal value of DDT and later synthetic insecticides became clear, it was not surprising that these chemicals were rapidly seized as easy solutions for old problems. What is surprising is how quickly entomologists seemingly forgot that pest control is a biological, not a chemical, problem. In the postwar years there has been an enormous emphasis on chemical technology disguised under various biological headings. Most practicing entomologists came to feel that the solution to all insect control problems lay in·chemicals. All one had to do was select the proper chemical and apply where needed. There was, of course, opposition to inordinate dependence on chemical controls (as examples, the early warnings of Nicholson, 1939, and Wigglesworth, 1945), but in the flush of hopeful enthusiasm for final solutions the biologists who held the opposing view were not seriously regarded. Now we see that the opposition correctly appraised the biological environment. Biotic complications following chemical control soon became as serious as the problems they replaced, or more so. Even now some applied biologists are reluctant to concede that chemical control practices need changing. Fortunately, their voices are fading.

Present approval of integrated-control systems stems largely from the overenthusiastic use of chemicals in pest control. The problems discussed in this book remind us that chemical pesticides are not the only means of controlling pests. Nor is there any single method. Basic dependence on natural controlling agents must be appreciated and fostered. Where chemical or other controls can be used to assist natural controls, they should certainly be called upon. But the pest control practitioner must first be an ecologist, whatever other tools or disciplines he may depend on to assist him in his ecological manipulations. The integrated-control concept puts perspective back into insect control. It has been lacking for the last two decades.

Integrated control depends on procedures that can be, and normally are, used separately. There are some more or less minor procedural characteristics that set it apart from other approaches to control, but, basically, its methods are borrowed from the traditional approaches. Integrated control is, in fact, not so much a procedure as an attitude, an outlook on how best to control pests. Properly practiced integrated control, as now in effect in such widely diverse crop environments as

Nova Scotia apple orchards and California alfalfa fields, accomplishes these things:

(1) Chemical applications are less frequent and are often in smaller amounts, leading to lower costs to growers and reduced hazards to both growers and consumers.

(2) Reduced chemical use leads to fewer biological complications Among these are lesser likelihood that resistance will develop and a greater survival among natural enemies of pest species.

(3) Reduced chemical use enables successful biological manipulation of two kinds: rebuilding of stable biotic complexes previously simplified by pesticides, and the introduction and survival of imported biological-control agents specifically aimed at target pest species. Both means, by increasing the complexity of fauna, result in greater stability and fewer numbers of all species.

The proper integration of control techniques requires a great deal of knowledge. It has become standard practice to depend upon the most knowledgeable—the expert—for advice in pest control matters. Advisers and consultants in production biology are now legion. But more than their occasional counsel is required. Integrated control demands knowledge, inspection, and responsibility. These requirements lead to much closer supervision of pest control. Supervised pest control has been practiced in German forests for 25 years (Franz, 1961). With perhaps less rigidity than in Germany, supervised pest control is the rule in American forests. To a lesser extent this applies to pest control in orchards. In both orchard and forest, long tree life and relatively great biotic stability make supervision and integrated control feasible. Only in very recent years has the feasibility of integrated control in annual crops been demonstrated (Stern *et al.*, 1959).

Here are two recent examples to show that integrated pest control schemes can indeed be successful in annual crops.

The spotted alfalfa aphid (*Therioaphis maculata*) appeared in California in 1954 and soon spread throughout much of the alfalfa-growing regions (Stern and van den Bosch, 1959). Parathion and malathion were immediately put into use to reduce aphid numbers. But the control was temporary at best, and the amounts and frequency of chemical application had to be increased. Resistance among the aphids to these insecticides began to appear, in combination with near-total destruction of the natural enemies of aphids. The result was phenomenal numbers of aphids and unprecedented crop damage. The inherent danger of total reliance on insecticides in aphid control became clear. A return had to be made to a better degree of natural control. In many trials, a systemic insecticide (Systox) reduced aphid populations, while harming natural enemies very little. Moreover, insecticide-induced mortality did not have to approach 100 per cent for satisfactory control. Lower doses could be used. Meanwhile new enemies—insect parasites and fungi—were successfully introduced. The combined actions of a selective insecticide and natural enemies, both native and introduced, now keep the aphid populations in satisfactory check without excessive cost, great toxic hazard, or development of insecticide-resistant populations. In this instance an occasional dependence on a toxic chemical serves only to complement natural controls, not to hamper them.

The second example concerns an established pest and a more complicated series of adjustments against it. Hornworms (*Protoparce* sp.) on tobacco are usually reduced in numbers by insecticides. Most states in which hornworms occur also recommend cultural control by burning or plowing-under the tobacco stalks that remain after harvest. Hand-picking of larvae is sometimes recommended. Generally, there has been little effort to control them with natural antagonists, but any successful means of doing so would bring at once the advantages of lower cost of control and reduced residue problems (a particularly important aspect of insecticide use on tobacco). In recent years Rabb and his colleagues at North Carolina State College have successfully integrated a variety of control techniques against hornworms that are remarkable for their effective simplicity (see, for example, Lawson *et al.*, 1961). Paper wasps (*Polistes* sp.), known to be effective predators of horn-worms, were encouraged to multiply by providing easily constructed, artificial shelters around tobacco fields. (Increases in the bird populations of forests in Germany, England, and Russia have been achieved in a directly comparable way for similar purposes!) These shelters could be permanently stationed about fields or moved as needed, in the manner that beekeepers move hives for pollination. The predation of wasps alone accounted for a removal of about 60 per cent of the total population of hornworms. Under some circumstances reduction was greater. Occasionally, for a variety of cultural and climatic reasons, predator populations were low and insecticides had to be used to effect control. In these instances the chlorinated hydrocarbon insecticides TDE and endrin were used to reduce numbers of hornworms to levels at which wasps could assume control. In contrast to previous insecticide applications, chemicals were applied only to crown portions of plants, and then in lesser amounts, and fewer applications were required. Costs and residue hazards were reduced by this integrated scheme of biological, cultural, and insecticidal control. The net effect from any viewpoint is an improvement on prior control methods.

The future of the integrated-control concept is excellent. It rests upon the awareness of applied biologists that crop ecosystems, like all others, are complex living communities. Insecticides intruded into them inevitably cause disturbances (Pimentel, 1961). In view of the serious nature of these disturbances, single-minded attention to target species cannot now be condoned. When an ecological basis for insect control is widely appreciated among entomologists, insecticides can properly be used to correct the ecological imbalances that occasionally must occur in crop ecosystems. The entomologist will then become an ecological counselor—like the physician, a prescriber of chemicals. "Prescription entomology" is a proper way of describing coming patterns of insect control.

4 · LOSS, COST, AND GAIN

The need to control pests of a great variety of kinds is obvious and universal. While there is no need to justify a greater advantage to ourselves in biological competition, we must be sure that advantage will in fact accrue when pest control is undertaken. The public acceptance of chemical control programs has rested upon an assumption of general advantage. The first portion of this chapter describes the goals and manner of making judgments in three categories of use: The first is in public health; the second, in crop-protection—agriculture and forestry; and the third, a miscellany of limited but well-defined uses that I term "special." All three are use categories. There must be a source of the chemicals put to the foregoing uses; the production and distribution of chemicals are undertaken by the large and rapidly growing agricultural chemicals industry and its ancillary services. Therefore, the fourth and last section of this chapter briefly describes the amounts, variety, and distribution of this industry's products.

Pest Control in Public Health

When the cause of discomfort, debility, or death of man is an animal that carries an infectious microorganism or a parasite, we condemn the carrier animal as well as its offending passenger. Such an animal is a vector, and most public-health pest control programs are vector control programs. By custom, nuisance insects are included under vector control.

Anopheles mosquitoes carry the sporozoan parasite that causes malaria. Since we cannot deal effectively with the malarial organism by itself, we aim at its carrier. Mosquito control is now conducted on a large scale throughout the world. The social consequences of malaria reduction and eradication are beyond measure. Direct control of the disease results, of course, but other benefits are an improved vitality that gives rise to a more productive people and one more resistant to other diseases. Knipling estimated in 1953 that no less than 5 million lives had been saved and over 100 million illnesses prevented through the use of DDT in controlling insect-borne diseases. Many millions could now be added to these figures. Death rates in Madagascar and Ceylon were almost halved within two years of the introduction of DDT for mosquito control (Hess, 1956).

More than a dozen other important diseases are transmitted by mosquitoes. These include yellow fever, encephalitis, filariasis, and dengue fever. Taken as a group, there is probably no more serious set of diseases to which mankind is exposed. And there are other vector-borne diseases, such as tularemia and scrub typhus, that, at least locally, take their toll. No region of earth is totally exempt. In the United States the most serious mosquito-borne disease at present is encephalitis in its various forms. The incidence of the disease is relatively low and is kept low by continual source reduction, sanitation, and prevention of access of mosquitoes. The amount of time and expense which goes into the study and prevention of encephalitis is justified by the results.

A massive effort has gone into the attempt to rid Central and South Africa of the tsetse fly, *Glossina,* a carrier of the trypanosome protozoan causing sleeping sickness. Locally—that is, in terms of a few thousands of acres—areas have been made habitable for man and his domestic animals. Here the reward of successful control is not measurable. The difference between success and failure is the difference between permanent productive human occupancy of land and a transient tenancy of the least productive kind. Large areas of Africa still remain to be cleared of the vector.

With such high goals in mind, one may wonder little about the price. In this instance tsetse fly alleviation has employed all the classical means of control—direct reduction of numbers, sanitation by removing favored kinds of brush species, and slaughter of wild game that act as a reservoir for the trypanosome. Hundreds of thousands of head of big game have been killed to satisfy this last requirement. Chemical means are also effective. In one area in Nigeria where tsetse flies concentrate during the dry season, treatment of forest islands with DDT at the rate of 20 pounds per acre eradicated the fly from seven square miles. The cost of this operation was officially listed as $2,000.

Perhaps all these methods seem justifiable in relation to their purpose. The total price of tsetse fly control is known to be high, but the intangible costs are incalculable. The value of the wildlife slaughtered cannot be calculated. The total effects of massive chemical applications over very large areas are not known. The biotic, recreational, and aesthetic worth of nature is beyond all tangible assessment. Also beyond measure, we must conclude, is the value of human life.

Pest Control in Agriculture and Range Management

Crop losses from insects and other biological competitors are unquestionably important. What value we may assign to their alleviation is not always clear. But the extent of loss is now at least partly measurable.

Only in the recent era of emphasis on productive efficiency and high marketing standards have partial crop losses seemed so important. In 1891 Fletcher estimated annual crop losses to insects in the United States to be 10 per cent of yield. For many years, 10 per cent loss was accepted as a "rule of thumb" estimate on aggregate tax by insects. The losses, of course, varied from crop to crop and year to year. Other kinds of crop losses from pests were quite probably double the loss from insects.

Decker (1954) suggests three major corrections that must now be applied to Fletcher's 10 per cent rule. The first correction is to the 10 per cent; at the time it was probably an underestimate. Secondly, many of the most destructive insect pests were introduced from other countries about 1891 and shortly thereafter; and, lastly, correcting for the present, American market standards have been raised so that insect-damaged produce can no longer move in open commerce. These statements imply a continuing change in the kinds of insects that cause damage and in the nature of damage. They further imply that the estimation of loss is strongly conditioned by the size of the fraction of the total yield for which the ultimate consumer will pay. Neither pests nor standards remain the same.

Losses to pests are usually based on "before-and-after" estimates, or, less often, actual yield figures. Yield changes may be translated into dollar gain or loss, and hence provide the only good measure of the worth of a pest control procedure. Even at that, they do not represent the only measures of worth that can be applied.

Convincing data on losses to pests are difficult to find (well illustrated by Parker, 1942). The general estimate is the least reliable. Loss from rodents in the United States, for example, is estimated at one to two billion dollars annually. The loss to insects is adjudged considerably larger, perhaps four billion dollars. The total estimate of the cost of weeds has ranged to eleven billion dollars. Figures of this type are largely meaningless. They may be offered to a legislature in support of increased budgets, or to the public in a quest for sympathy. For a given program estimates are rarely analyzed regarding how the figures were obtained, how certain the benefits of crop protection are, or, for that matter, what force alleviation might bring to bear on existing production and distributing systems. And the figures have in addition no real impact on an audience already aware that pests are pests and cause a very great deal of trouble.

Specific estimates applying to particular crop types are more useful. The best summary of these is contained in a publication of the Agricultural Research Service (1954). In this summary the difficulties in preparing estimates are illustrated repeatedly by direct statement or by implication. The indecisiveness of loss figures brought together in this report by the best qualified scientists and economists in government led to the recommendation for further studies on the problem. A National Research Council committee has since been appointed to study the question, but no reports have been issued.

What is meant by loss must first be made clear. It is one thing to recognize that losses occur, another to do something about them, and still another to determine whether the alleviation of loss is necessarily socially advantageous.

Losses in agriculture are of two types: one involving a commodity, the other involving the productive base on which commodity production depends. Any reduction in quantity or deterioration in quality during the production, handling, and processing of farm and forest products is commodity loss. Deterioration in land on farms or forests is a productive loss. Losses of either type are not always preventable. Some are and will remain unpreventable. Some are presumably preventable, though not under the existing economic climate. Some are preventable under current social and economic conditions. The last category is the only important one.

Partial alleviation of loss is possible, but would not approach total relief even if all pest control technology were put into use. Current losses in many crops could be substantially reduced if present knowledge were fully employed. Losses of cotton, for example, could be reduced 75 per cent if "total" farming—use of resistant varieties, use of mineral supplements, and the like—were practiced (Agr. Research Serv., 1954). (In this time of acreage restrictions, production quotas, and supported prices, would this be a wise thing to do?) Inertia, lack of education, and economic factors preclude any such productive efficiency. While examining specific loss figures, therefore, one must constantly keep in mind that loss estimates are total but that alleviation of loss is partial. And further, partial alleviation very rarely achieves its full potential.

There are two general ways by which crop loss from diseases and pests, and gains from attempts to alleviate loss, can be described. The first method, older and more general, employs percentage change from a presumably realizable total production in a single commodity. The second also deals with single commodities but depends on productive units, either change in yields per acre or their economic mirror, dollar gains.

Two commodities—cotton and apples—are used here to illustrate the applicability of the first method. Note that the emphasis is negative, that little recognition is given to the extent to which total loss is correctible. Note also that only producer standards enter the accounting.

Cotton insects account for an annual loss estimated at one-fifth billion dollars (Agr. Research Serv., 1954). The boll weevil accounts for about 10 per cent of the annual cotton crop loss. Of course, there are regional and yearly variations; in some years the damage is very great. To this loss must be added the cost of control, which itself varies considerably. In 1952, a year of light damage, the cost of control for boll weevil alone was calculated at $75 million. All other cotton insects cause damage about equal to that of the boll weevil—about 10 per cent, or a dollar value of $106 million. Here, too, the cost of control must be added.

In 1940–44, one of the best-known fruit pests, the codling moth, caused an annual loss of 15 per cent of American apple crops, or about $25 million (Haeussler, 1952). The cost of control was perhaps the same. In 1944 to 1948 the annual loss was about 4 per cent, or about $9 million dollars. The marked reduction in loss was due almost entirely to widespread use of DDT. (Insecticide resistance and rise in spider mites as pests, because of DDT, might be charged against the DDT gains, however.)

The second method is a better way of representing loss figures than is total crop assessment. The National Agricultural Chemicals Association has compiled data on the economic benefits of pesticides and presented them in tabular form. Table 3 provides some selected examples to illustrate both yield and dollar gains.

TABLE 3

A few examples of yield gains and net dollar gains from pesticide use in common crop species (from National Agricultural Chemicals Association, 1957)

Crop and unit	Pest	Yield/acre untreated crop	Yield/acre treated crop	Treatment cost/acre	Net gain/acre	Source
Barley (bushel)	Weeds	45.5	49.5	—	$5 (gross)	N. Dak. Ext. Serv.
Seed cotton (lb.)	Bollworm	7203	7860	$16	126.50	Texas Agr. Expt. Sta.
Pea seed (lb.)	Fungi	456	610	.70 (seed treatment)	21.25	Boyce-Thompson Inst.
Tomatoes (ton)	Diseases	5.4	11.8	40	100.00	T. N. Jarrell Co. (Md.)
Sugar beets (ton)	Root maggot	11.8	14.5	2.25	32.20	N. Dak. Agr. Expt. Sta.

It is clear enough from these selected examples that both yield and dollar gains can result from pesticide use. A certain caution must be observed with such sets of figures. The means of arriving at such values are not very refined and are often unreliable. Moreover, they tell only part of a story. Too frequently, they reflect an unrealistic "optimism" not in keeping with their source.

We may next ask, what is the actual expenditure for pesticides to achieve gains in yield? To answer this, I will depend on agriculture in California, which is both intensive and chemicalized; a large part (some 25 per cent) of all pesticides used in the United States are used there. Probably the most reliable data on pesticide worth to agriculture come from California. C. A. McCorkle, of the University of California, recently (1959) prepared a summary illustrating the relative importance of chemicals in producing California's high-yield crops. Table 4 describes relative cost inputs in certain selected crops in the year 1957.

TABLE 4

The costs of pesticides in relation to total production cost in selected California crops (from McCorkle, 1959)

Crop	Pest and disease control	Total cost to harvest/acre	% pesticide cost
Cotton	$10	$175	5.7
Rice	4	75	5.3
Alfalfa	5	85	5.8
Peaches	35	400	8.7
Pears	40	315	12.7
Tomatoes	7	248	2.8

Both growers and advisers are convinced that these cost increments, ranging here from 3 to 13 per cent, are essential to the production of high-quality, high-yield produce. In areas where farming is less intense, many growers do not use pesticides at all. In the early 1950's in Iowa, for example, only 14 per cent of all growers used any variety of pesticide on crops (and yet we think of Iowa as one of our most productive states!).

One may wonder about the accuracy of specific loss figures. In 1953 the rice leaf miner outbreak was credited with ruining 10 to 20 per cent of the California rice crop. Judged by my personal damage appraisal at the time of the outbreak, this range was certainly correct for many of the rice fields. Even at that time, however, by far the greater part of the rice acreage was not hit at all. To the presumed loss of yield must be added a cost of some $1.25 million for pesticides (and a very large wildlife loss). Curiously, however, rice yields that year were among the highest on record to that date (Table 5). Equally inexplicable was the reduced yield per acre and consequently reduced total production that occurred with much greater acreages in 1954, the year following the outbreak and treatment. The price of rice was also substantially lower in the year following the outbreak. I conclude, therefore, that growers extrapolated single-field losses to the total acreage, and that "insurance" spraying was the rule. Economically, insect damage seems relatively much less important than other factors. One may perhaps be permitted to wonder if perspective as well as accuracy were not somehow awry in reporting of the "threat" to the rice crop.

TABLE 5

California rice production and value in some recent years. Data from annual reports of the California Department of Agriculture.

	Acreage harvested (1000's acres)	Production (1000's bags)	Average yield per acre (100 lbs.)	Value (price/bag)
1951	314	10,676	34	$4.95
1952	330	11,880	36	5.95
1953	412	12,257	30	5.38
1954	453	10,872	24	4.48
1956	286	12,012	42	4.44

In all the foregoing examples it is clear that losses to insects may be economically significant; it is also clear that steps to correct loss are at best only partly successful and on occasion are not convincingly advantageous.

In bird and mammal control operations, not only must all these qualifications be made but we must also add intangible elements. Depending on the animal in question, these elements might be aesthetic appeal, sympathy, fear, or repulsion. There is no species of competing bird or mammal that does not have its defenders. Eradication of a vertebrate species is never an accepted goal, and "control" is best described as local reduction in numbers.

Bird depredations are particularly difficult to appraise accurately. Some alleged depredations better rate as nuisance. Some allegations of crop loss are naive; most overlook the value of birds as destroyers of insects and weed seeds. Very rarely are serious attempts made at cost analysis. Calm attempts to analyze the proper worth of bird depredations are as rare among control personnel as are concessions among bird-lovers that damage is possible. Analysis of this kind is difficult at best, and is impossible without integration of biological and economic information with aesthetic judgments.

The best recent attempt to my knowledge is the study of blackbird depredations by Neff and Meanley (1957), both of whom approach the analysis with knowledge of production problems, tempered with the sensitivities of those who love wildlife. Throughout the United States are many species of blackbird that congregate in large numbers, either in nesting colonies or in fall and winter movements. Large flocks are responsible for a variety of loss allegations—eating rice heads "in the milk" in early season, shattering drying heads of milo maize immediately prior to harvest, breaking down stems of cereal grains, and so on. Neff and Meanley conclude that serious losses do occur to individual rice growers, even to the extent of one-half or more of the crop, but that such losses are rare. The average farmer suffers little if at all. Based on farmer estimates (which normally describe *worst* damage as if it occurs in all fields), these workers calculate that 1.28 bushels per acre is the annual loss of rice to birds in Arkansas. In rough figures the state's loss is $0.5 to $1.5 million. Such estimates do not take into account estimates of the birds' worth in insect control. Methods of reducing numbers or of preventing access of birds are known and, if applied, would greatly reduce loss. Most growers do not consider the problem serious enough to merit their consistent attention.

Because of the mobility of birds, damage caused by them is apt to be spotty and unpredictable. Reasonable effort can prevent most serious damage, even if birds do appear in numbers. However, there is in Africa one species—*Quelea quelea*, the Sudan dioch—that damages crops on a staggering scale. So severe are its depredations that entire crop development schemes have had to be abandoned though the areas were otherwise suited. Plantations of cereal grains and surface water availability through irrigation have stabilized the bird's movements while increasing its numbers. Heroic effort, through poisons, explosives, and flame throwers, has been expended to reduce its numbers, as yet with only local success. For example, in the Senegal Valley 30 million birds and 46 million nests were destroyed in 1955 (Heseltine, 1960). Heseltine insists that *Quelea* is a far more serious threat to agriculture than the locust, whose plagues in arid Africa are legendary. He suggests establishment of an international commission to study and to develop methods of control, after the nature of the Anti-Locust Commission, which has been in operation over thirty years. No other bird can lay claim to such importance in agricultural depredations.

Predatory, nuisance, or disease-carrying mammals have few defenders. There is, accordingly, rather little general concern for their protection, and equally little concern for proper estimation of their worth. Rodent control is the main effort, and until recently has been largely concerned with proper sanitation and disease prevention in urban areas. Intensified crop production has in recent years focused attention on competing rodents in field, forest, and rangeland. Next in importance are carnivorous forms, such as coyotes and foxes, which kill and harass livestock and poultry. Some of the smaller carnivores have recently become important as disease carriers, notably of rabies. Of lesser importance are the ungulates—deer, goats, and antelope—which cause local plant damage or compete with livestock for forage.

The control of rodents and carnivores is incited further by a traditional attitude of fear and repulsion. Hence, locally at least, reduction of numbers does not require justification to most people. Even so, the rationale of control is often found wanting when closely examined. A desire to preserve rodents and carnivores does not normally form the core of objection to control operations; it is the fact that mammal control methods in the field are neither tidy nor specific. They depend on trapping or poisoned baits. Both methods are nonselective. Desirable species of mammals—furbearers, livestock, and pets—are too frequently the unintended victims (Rudd and Genelly, 1956).

Game species—the ungulates and the large carnivores—are accorded a more sympathetic tradition. A conflict in values is usual when these species become pests or are identified as disease carriers. Although the resolution of different attitudes toward them may be difficult, their control is not. Reduction in numbers is selective, depending chiefly on guns. Repulsion, whose purpose is to prevent damage without reduction in population number, is sometimes practiced locally. It depends chiefly on frightening devices (explosions, etc.), but in recent years chemicals effective for this purpose (but expensive) have become available. With either approach the result is selective. Trapping or baiting, used rarely, normally exploits peculiar habits of the target species so as to render the technique as selective as possible.

Close analysis of the costs and benefits of mammal control is rare. In most instances the value received, although believed to be obvious, is reckoned intuitively and subjectively. Less commonly, the value received is clear enough to one faction but not accepted by others. I have selected the following examples to illustrate why mammal control is undertaken. These examples range from generally acceptable purposes to those acceptable only to certain factions.

Commensal rats and mice, which produce filth, contaminate foodstuffs, and carry disease, receive no defense. Enforced high standards of sanitation would do away with them as a serious problem. Since acceptable uniform standards of sanitation are not socially achievable, urban rats and mice must be attacked with trapping and with poisoned baits. No justification for this kind of control is needed. Reference should be made to Zinnser's *Rats, Lice and History* for a fascinating account of man's historical relationship with commensal rodents.

Only on croplands can one properly weigh economic value without subjective complications. The living crop complex is purposefully unstable; any mammal that removes or disturbs an economically significant fraction of a crop is considered by the grower as a fair target for control measures. With ever-increasing efficiency in farming, the allowable size of this fraction continues to decline. Vole (*Microtus*) populations in irrigated pasture or on alfalfa hay fields will illustrate. Normally, only one species of plant is present, and this monoculture is carefully irrigated and cultivated to produce maximum yields of vegetative parts. A vegetative feeder such as the vole, easily able to dig in moist soil and sheltered with a dense plant cover, could not find better conditions for survival and increase. It is not unusual to find a "standing crop" of 100 to 150 mice per acre. Jameson (1958) showed that captive meadow mice each ate an average of 9.8 grams of alfalfa (dry weight) daily. In an acre of alfalfa 100 mice would consume about 4 per cent of an annual production of 11 tons. Or, on an acre of sheep pasture, 100 mice would consume 12 to 13 per cent as much forage as 10 growing lambs. An ecologist would question whether the accounting can be kept so simple, but a crop producer would not. This is the best kind of data he has on which to make control decisions. If he believes he can "save" perhaps one-half ton of forage worth 10 to 15 dollars, he will be ready to spend 2 to 4 dollars to do so. When the population of mice exceeds 100 per acre, a field will clearly show the increase in numbers by decreases in the size and density of forage plants. A decision to control is easily made under such circumstances. On the other hand, densities below 50 mice per acre are by far the more common. Static populations at no greater density are not worth controlling.

The control of a predatory mammal requires a more complicated rationale. The coyote is the chief target of present predatory-animal control efforts. That losses to individual livestock growers from coyotes are sometimes severe is beyond question. But this fact alone cannot be considered the overriding element in the decision by public agencies to initiate control campaigns. The issue of coyote control is clouded by charge and countercharge, a good deal of poor biology, and the probability that current control programs administered by state and federal agencies cannot be economically justified. This is the question: Is the cost of the predatory-mammal control campaigns of the U.S. Fish and Wildlife Service and cooperators justified by the losses sustained by growers and ranchers? Practically, the question is not subject to a single answer. We may choose to look upon predatory-animal control expenditures simply as one of the first of the now-common federal subsidies to agricultural producers. Some growers are probably helped; most are not. The livestock and poultry industries are convinced that the program is essential and maintain effective lobbies to pressure for continuation of federally sponsored control operations. But only the occasional grower who has sustained serious loss actually gains. Costs of the program are met largely by taxes. In some areas cooperating livestock grower associations assess themselves on a per head basis, and state and county game funds are often diverted to support predator control. Locally, public-health agencies may contribute to such interests under the guise of vector control programs. These are the sources of income. Money is spent chiefly in the western states, and largely on public lands. The issue is strongly influenced by regional legislatures, and the question of the use of public lands for widest public benefit is normally avoided.

If the question of predatory-mammal control is to be argued solely on economic grounds, one has the right to expect of control officials that loss, cost, and gain will be properly studied, documented, and presented to legislators. Yet agencies responsible for control have never felt it necessary to do these things (see Chapter 16). Several competent reviewers have attempted to do what the federal agencies should be expected to do but have not yet done. The detailed accounts of Bond,

Hall, and F. C. Craighead, from which I have taken the following data, give the best cost-accounting available.

Bond (1939), for example, made a careful study of coyote food habits at the Lava Beds National Monument. In one year only 12 sheep (0.1 per cent of the population) were killed by coyotes (2 per section). Largely responsible for the low sheep loss was the simple expedient of using road flares to keep the the coyotes at a distance from the flocks at night. Control workers reported that the expenditure for killing a single coyote was $1.40. Even at this conservative figure, which is exceeded many times over in other parts of the state, the local sheepmen could hardly justify killing on a strictly economic basis if they had to pay for it themselves. The program is accepted because it is done "free" by the government.

Hall (1946) has examined the reports of the Nevada State Rabies Commission. During an 18½-year period almost a million dollars was expended to destroy 88,601 coyotes, 14,352 bobcats, 56 mountain lions, 4959 badgers (first 5 years), and 4509 other furbearers (including badgers). Over $104,000, realized on the sale of marketable fur, was deducted from the total cost. Coyote control was the object of the commission. This being so, the cost of each coyote averages $9.93, and for all "predators," $8.54. A rabies epizootic had initiated the formation of the commission. The disease is, of course, terminal, and its effect on the coyote population was profound. Within a short time the commission had expediently changed its function from rabies suppression to the reduction of livestock depredations.

F. C. Craighead (1951) has provided the best general evaluation of coyote predation I have seen. His account should be read to appreciate the full range of the problem. There are still many unknowns in the equation, but these conclusions can be made. Allegations that coyotes seriously affect big-game populations border on nonsense. In fact, coyotes serve important functions in scavenging and biological sanitation. Carrion is regularly eaten, and "living" carrion (weakened or diseased animals) is removed from the population to the advantage of survivors. The critical question is whether coyote damage to livestock outweighs the service that coyotes perform in removing rodents that compete for livestock forage. Both rodents and coyotes can do a great deal of damage. Almost three million coyotes were taken in the United States in the thirty-year period 1916–46. In subsequent years the use of "coyote-getters" and poisoned bait stations has made it impossible to count the "take," but it is generally conceded that these newer methods of control are more effective than prior methods. The Predator and Rodent Control Division of the U.S. Fish and Wildlife Service has had a budget between two and three million dollars annually for many years. It is not possible with published information to calculate precisely how much it has cost to control the perhaps six million coyotes destroyed to date. It is known, however, that rabbits and rodents form by far the greater portion of the coyote's diet. Decreased predator control would reduce the need to conduct the costly range rodent control campaigns initiated in the last few years. The proper use of rangeland is actually the key to controlling predation losses (Chapters 15 and 16). Balanced use ensures fewer rodents, lower costs in both predator and rodent control, and better sustained livestock grazing. To these tangible gains must be added the wildlife hazards avoided by reduced control programs and the preservation in numbers of a picturesque member of rangeland fauna.

Large-scale federally supported predatory-animal control began during World War I. In 45 years the control agencies have not provided a thorough, careful analysis of its costs and advantages. Such an analysis does not seem to be requisite to continued federal support of the program. Presnall (1949: p. 589), in a particularly forceful article, described the attitudes that in my experience most characterize the predatory-animal control advocate. A few quotations will illustrate. Speaking of the western range:

> Other esthetic values of wildlife, several of which are intangible, might cause the assumption that game (esthetic value) is to livestock (economic value) as 1 is to 10. However, since esthetic values are largely intangible, no exact numerical comparison is possible. It can be said only that livestock values outweigh wildlife values. This provides a yardstick for determining policies of land management, of which predator and rodent control is one phase. It is admittedly an imperfect yardstick. It does not consider the possibilities of free tariff in relation to survival of domestic livestock production; or the likelihood that synthetic fibers may largely supplant wool; or the sociological and ethical implications of possible conversion of the West into a vast national picnic ground where a few cattle and sheep might be tolerated for their value in flavoring the yeast-brewed synthetic protein foods of the future.
>
> It is a practical, present-day yardstick. . . .

The "yardstick" suffers from some real and most tangible omissions, some of which I have already described. I can do no better than to quote a portion of T. I. Storer's commentary at the conclusion of Presnall's paper (p. 591). "We have some good information on food habits. We have a little information on the amount of economically valuable material taken by rodents and predators. We scarcely have any real evaluation in a quantitative or monetary sense of the damage done, particularly by predators. Those facts are still to be obtained and we cannot view this matter entirely in an objective manner simply because we do not have such information at the present time."

Storer made the foregoing statement in 1948. There is no need to change it now.

When a mammal pest is also an important game species, the assessment of need for its control is difficult to render and, when made, will assuredly not be acceptable to all. Deer are the best-known pests in this category. They enter orchards to eat new growth; they browse freely in irrigated pasture and alfalfa; they eat rangeland forage that might go into livestock; they interfere with reforestation projects; and they frequently damage row crops such as beans when these are near deer ranges. Occasionally, their depredations are serious.

Fences and repellents are commonly depended upon to prevent access to a vulnerable crop. So also are frightening devices—lights, exploding devices, and patrols. But most common is controlled killing. Dependence is placed on hunter kill during regular season, but other means can be used to reduce deer numbers locally.

Every management device recognizes the value of deer as game. Acknowledgment of their liability came late, but that too is now widely recognized. In New Zealand the liability seems to be much greater. There are several species of deer, all of them introduced. Also introduced are several other grazing and browsing mammals—opossums, kangaroos, rabbits, hares, and, of course, sheep and goats. The economy of New Zealand is based on sheep. Therefore any serious competition for forage from deer or other undomesticated herbivores is a matter for national concern. Graziers and deer-stalkers have argued on this conflict for many years. More recently, the charge against deer has become more serious. New Zealand is geologically young; the topography is rugged, the soil thin. Rainfall is high, and so are erosion rates. The rate of erosion has recently increased markedly. Forests are changing in appearance, and watersheds seem to be losing their ability to retain water. The changes are apparently profound. Foresters and agriculturalists give deer a good share of the responsibility for these changes. Deer-stalkers and some ecologists challenge the assessment and maintain that damage is overrated or the result of natural processes. Damage by sheep is also rated in some sections, and should be acknowledged for what it is, not charged to deer. The deer-stalkers further maintain that where local abundance of deer produces problems, numbers can be effectively reduced by hunting.

Graziers, of course, are profoundly influential. Without serious study, therefore, it was a simple matter to arrange for large-scale deer "eradication" programs. The ultimate technique used was aerial dissemination of 1080-treated carrot baits. Compound 1080 is distinguished by its high toxicity and stability, thereby making secondary poisoning a likelihood. Not surprisingly, both deer-stalkers and lovers of the native fauna became alarmed.

The controversy seems distinguished by an absence of convincing data. Nonetheless, it had become so important that in 1958 a national meeting was called for its resolution (New Zealand Forest Serv., 1958). The meeting was attended by several ministers of departments and members of parliament. Agreement was not forthcoming. Possibly it is accurate to say only that the position of the foresters and graziers was politically solidified. The control program not only has continued but has intensified. There are no reports available describing the effects of herd reduction on forest or watershed, the impact of deer removal on other herbivores, or the effect of poisoning on the native fauna.

Pest Control in Forestry

Much of what has been said of agriculture applies to forestry. But there are some important qualifications to make. The forester is not faced with an annual crop. Trees take many years to mature, and the value of the annual increment of growth is lower than in most agricultural crops. Protective measures must be correspondingly lower in cost. Trees are able to withstand more insect-caused damage than crop plants. Moreover, forests are more complex ecologically than croplands. Foresters and entomologists therefore prefer to exploit natural controls where possible.

Serious losses do occur in American forests. The chief sources of damage are fire, disease, weather, insects, and mammals. Damage is not always preventable or controllable. Irrespective of cause, losses derive from mortality of trees, increase in culling rate, and reduced growth rate. Insects and disease account for large losses in all these categories, but the amounts and significance of loss are difficult to estimate. The U.S. Forest Service (1958) estimates insect-caused mortality losses of saw timber to be 5 billion board-feet annually. Mortality loss from disease was estimated at less than half that amount, 2.3 billion board feet. Growth losses caused by disease and insects were believed to be far larger than mortality losses. Diseases were responsible for an estimated annual growth loss of 17.6 billion board feet, whereas insects were adjudged to cause only 3.6 billion board feet of growth loss.

Direct disease control depends on silvicultural practices—eliminating carrier plants and adjusting age structure, for example. There are no new and revolutionary weapons in disease control such as one witnesses in insect control. Yet almost $4 million was spent on disease control in the United States in 1952, most of which went to efforts at control of a single disease, white pine blister rust. In the same year direct forest-insect control expenditures by state, private, and federal agencies totaled $3.6 million. Gypsy moth control was nearly half the cost, with the Engelmann spruce beetle and the spruce budworm accounting for most of the remainder.

TABLE 6

Estimated timber losses caused by six major outbreaks of forest insects in the United States (from U.S. Forest Service, 1958)

Cause	States	Date	Approx. volumes of killed timber trees (millions of bd. ft.) *
Spruce budworm	New England	1910–19	8,000
Spruce budworm	Lake States	1913–26	5,800
Mountain pine beetle	Idaho–Montana	1911–35	15,000
Western pine beetle	Oregon	1921–37	12,600
Western pine beetle	California	1931–37	6,000
Engelmann spruce beetle	Colorado	1940–51	5,000

* Only 2 per cent of the total killed saw-timber volume was reported as salvaged.

Severe insect outbreaks produce losses of great dimensions. Very frequently, they are associated with natural catastrophes—fire and wind. The Forest Service identifies six major insect outbreaks in the continental United States from 1900 to 1952. These are listed in Table 6, with the estimated amounts of saw-timber lost.

The spruce budworm, the Engelmann spruce beetle, and the introduced gypsy moth are currently causing great concern and are being heavily treated.

In carefully managed forests—more particularly, monocultures—growth increment has become an important economic consideration. Traditionally, and still the case in northern mountainous forests, severity of damage of insect and disease depredations has been gauged by mortality of saw timber. This method of assessing severity of depredations applies particularly to the forests in which the spruce budworm periodically explodes. The budworm outbreak in Maine, Ontario, and Quebec reached severe proportions in the middle 1930's, and has continued to spread. By 1944 the infestation covered 125 million acres in Quebec and Ontario. Mortality ranged up to 70 per cent in some areas. The area of irruption has since widened, although local populations fluctuate without correlation with others in the species range. (New Brunswick experienced heavy densities in the early 1950's.) DDT has been applied (sometimes repeatedly) over many millions of acres in efforts to limit the density of the budworm populations.

An important observation underlies all the foregoing examples. The greatest damage in forests comes from native (not introduced) insects. Instability in the numbers of forest pests, unlike agricultural pests, is the result of imbalances in natural factors. Also unlike damage in agriculture, forest damage is an occasional, not an annual, event. A rise in numbers of certain species is frequently predictable on croplands. Prediction is rarely possible in forestry. A search for the factors to permit prediction is the chief concern of the forest entomologist.

While the chief causes of loss in forestry are normally the native insects, in recent years intensified reforestation has been slowed somewhat by the depredations of vertebrate animals. Reforestation may depend on natural seeding or sprouting, on the sowing of seed, or on the planting of seedling trees. Natural regeneration has in some areas been prevented by mice, whose numbers increase greatly as the forest is "opened up." Normally, however, the failure of a tree species to re-establish itself by natural means should be taken as a sign of poor adaptability. Sowing of seed or planting of seedlings cannot be regarded as natural. These practices introduce a convenient food supply into an environment already well stocked with mice and deer. It is hardly surprising that such restoration efforts commonly fail. Barriers to access to seedlings, such as wire cones or chemical repellents on seeds, can keep down damage, but successful growth can be assured only by reducing the numbers of the offending animal. Locally, the losses of seeds and seedlings due to vertebrates can be serious. The solution is also local in scope. It requires sustained effort. There is as yet no easy method of ensuring against such losses.

Pest Control in Special Situations

The use of chemicals in limited but sharply defined situations is rapidly expanding. Chief among these uses is weed control. Control of herbaceous or shrubby vegetation along highway or utilities rights-of-way has become a multimillion-dollar enterprise (Egler, n.d.; Niering, 1958). The announced purposes are safety, "cleanliness," and convenience of maintenance. Without question the use on rights-of-way will expand. So also will use in those situations wherein safety is a more genuine concern—railroad yards, tank farms, refineries, and industrial plants.

Aquatic weed control is now widely practiced, and will assuredly expand. This expansion will come from sharply increasing recreational demands on impounded waters and from irrigation and other surface distribution of channeled water (George, 1960*b*).

Control of undesirable fish populations with chemicals is already a well-established practice. But it has sharply expanded in the last decade and will expand more as fisheries management intensifies.

The Production and Distribution of Pesticides

Few critics of pesticide use realize how vast and complicated is the industry that serves the pest control interests I have described. In this section, with the aid of a few production and use indicators, I wish to show how great the demands for chemicals have been and how these demands are apt to increase (see Gunther and Jeppson, 1960, for a point of view oriented toward the grower and industry).

In 1939, the industry showed about $40 million worth of sales. Twenty years later the figure was about $300 million. In 1975 the figure will have climbed to $1 billion (Fischer, 1956). Insecticides take the larger share of any figure, but whole new classes of chemical herbicides, defoliants, miticides, and others have altered the traditional components of pesticide value. These changes are largely due to altered farm technologies, differences in cropping systems, and corporate farm ownership. Wider acceptance of pesticides has been aided by public advertising—billboard, television, and newspaper. Selling to both agricultural consumers and the general public has in recent years become "big business," and has followed the advertising practices common to most commodity promotion.

DDT is the most widely used pesticide. It may be accurate to say that no other synthetic chemical has had such an impact on the world's population. Only a few thousand pounds were available in 1944, and this amount was an Allied military monopoly. Now DDT is produced in many countries, even in some with limited industrial capacities. The total amount produced annually must certainly approach one-quarter billion pounds. In the United States alone production of DDT in 1958 was over 145 million pounds. Ten years before, production of DDT was only slightly over 20 million pounds. About half of U.S. production is used in this country, the remainder going largely to Latin America, mainly Mexico. A few other countries also export large quantities of DDT. The bulk of this export comes from the United Kingdom, West Germany, and Switzerland. In general, exporting countries are also large users.

TABLE 7

Domestic consumption of DDT in 1952 (Arrington, 1956)

Country	Amount (metric tons)
France	1,500
Indonesia	about 1,250
Italy	under 850
Japan	about 2,000
Peru	about 140
United Kingdom	under 2,000
United States	about 23,000
Western Germany	about 3,000

Table 7 shows approximate domestic consumption of DDT in several countries in the peak year of 1952 (Arrington, 1956). The amounts largely reflect agricultural demand, but in two or three cases antimalarial campaigns are "hidden" in the totals. Household demands are rarely as much as 10 per cent in any country.

The basic sales price of DDT is about 50 cents per pound. In very large quantities its price might dip below 30 cents. The importance of

DDT to producers, in both volume and price, is clear. Most other synthetic chemicals are more costly; some are much more so. No pesticide is at the same time so cheaply produced and so widely used as DDT.

Most common pesticides were not marketed as few as 15 years ago. Table 8 compares the volumes of production of several of the commonest agricultural chemicals (Commodity Stabilization Serv., 1960). A table such as this can tell a good deal. For example, the production of lead arsenate—"the DDT of the inorganic pesticides"—declined sharply following wide-scale acceptance of DDT. However, production stabilized at a lower plateau, and it still merits being rated as a common insecticide. Once found widely useful, a chemical normally continues in some demand even though substitutes appear.

TABLE 8

Volumes of some common pesticides produced in the United States in selected years (from Commodity Stabilization Service, 1960)

Pesticide	Volume of production (1000's of lbs.)		
	1948	1953	1958
DDT	20,240	84,366	145,328
Parathion		2,999	5,439
Lead arsenate	23,000	14,196	14,938
2,4-D (all forms)		54,871	55,846
Combined—aldrin, toxaphene, dieldrin, endrin, chlordane, and heptachlor		29,000	98,280
Benzene hexachloride	18,400	57,363	30,797
Ziram		1,152	1,178

Note: Blank spaces indicate that no information was available or that the compound was not then marketed.

For comparative purposes one may say that blank spaces in the 1948 column mean that no significant amounts of any of these chemicals were available. The dramatic increase in production in the following five years illustrates the great change in the range of marketed offerings in agricultural chemicals. Similar patterns seem now to be occurring with other newly marketed chemicals. The great increases in the "combined" group of chlorinated hydrocarbons indicate a greater demand for toxic chemicals with a long residual action. The countering demand for a predictably short residual life leads to sharply increased consumption of organophosphorus insecticides, here illustrated by the most widely known member, parathion. The explosive increase in growth-regulating herbicides, exemplified in 2,4-D, drew attention to all aspects of growth manipulation. A few of the more refined usages are selective weed control, induced early flowering, and induced final fruit ripening.

Production figures are the best available indicators of the expanded use of pesticides. The few given in Table 8 clearly show a sharply widening demand in the postwar period. Projections from these figures show no declines in demand. Rather it seems probable that the rate of use will increase more rapidly as a rising world population places greater emphasis on high crop yields and as large-scale disease control and vector control campaigns touch more countries.

Area figures are less accurate than production figures as indicators of widening pesticide use because they do not tell what chemical is applied or how many times chemical treatment is repeated on the same acreage. Whereas cereal grains may be treated only once or not at all, orchard crops are sometimes treated a dozen times or more in a season. Cotton frequently requires weekly treatment. In fact, cotton probably requires more chemical attention than any other crop! But there is no "average"; hence, total acreage means very little. As a rule of thumb, chemically treated acreage in the United States approximately equals the acreage of intensively used farmland (about 400 million acres). Most crops now require some form of chemical treatment. To this figure must be added the acreages treated for forest and range insects, vector control, and nuisance abatement. These pesticide uses vary with apparent necessity. Although use for such purposes is unpredictable and opportunistic, it is also true that public control programs are now more widely accepted. Accordingly, there has been an increasing acreage treated for these varied purposes.

Aerial application of pesticides has particular interest here. The number of acres treated with pesticides by air is a reliable figure because the regulatory agencies control not only pesticides but aircraft. With the exception of intensively treated crops such as cotton and tomatoes, most applications are not repeated. Hence, treatment figures with appropriate qualifications give a generally accurate picture of single-application coverage.

Aircraft application as a method of pesticide application has grown dramatically. Moreover, many of the "opportunistic" uses alluded to above (particularly over range and forest) can be accomplished only with aircraft. The figures that follow give good relative data on the numbers of acres treated by air and the differential increase in aircraft as a method of applying pesticides.

In 1952, over 37 million acres were treated aerially in the United States. In 1957, the acreage covered had increased to over 61 million, and the dry and liquid pesticidal materials dispensed during that year were respectively 655 million pounds and 95 million gallons (Federal Aviation Agency, 1960).

California and Texas lead the states in the use of agricultural aircraft, California ranking first. Between one-quarter and one-third of all air usage by pest control operators in the United States is in California. Moreover, records of use are more accurate there and have been kept longer than in most states. In Table 9 is the approximate acreage treated with chemicals of any kind by air in California in the years indicated.

TABLE 9

Approximate acreage treated with chemicals by air in California, 1946–58. Data taken from Annual Reports, Bureau of Chemistry, California Department of Agriculture.

Year	Number of treated acres
1946	35,000
1951	3,000,000
1956	5,600,000
1957	5,320,000
1958	5,308,000

Of the total acreage treated by commercial pest control operators in California in 1957, about 81 per cent was treated by air. Spraying with liquids was 56 per cent of the total; dusting was 25 per cent. The individual grower normally handles his own ground spraying and dusting, but the commercial operator has been acquiring more of this trade.

The acreage chemically treated by aircraft in California has stabilized at a high figure, about 10 per cent of all acreage covered in the United States. Great expansion in use is not likely in California, but is probable in most other crop-growing areas. Greater emphasis is also to be expected in the coverage of noncroplands (range and forest) as new uses of aircraft in pest control are initiated. The aerial distribution in 1960 of 1080-treated rodent baits over 100,000 acres of California's forests illustrates such a new use.

So-called "mass" treatments of this kind will increase—and be paralleled by increases in environmental hazards. Treatment by aircraft is more economical than comparable treatment from the ground, but the "price" must fairly include a decreased ability to confine chemicals to target species or areas. Much research effort must now be expended to learn ways of countering this trend toward indiscriminate methods of application.

A Commentary on Probable Directions

This chapter was developed to give a glimpse of the different kinds and justifications of pest control. It is directed toward those outside of the producing or managing fields who have little awareness of the reality of pest control problems or of the volume of pesticides used in attempts to resolve them. I have found it difficult to present these data under single themes. How, for example, do we reconcile the paradox of our alarm at the loss of deer from 1080-treated rodent bait in California with the New Zealand government's deer "eradication" program using the harshest kind of baiting procedures? Many apparent contradictions render pointless any single justification for given pest control procedures. Conflicts of this kind are found throughout this text.

Complete agreement is not possible on what constitutes a pest or what procedures best suit its control. Hence, total judgments are normally not attempted if more than a single, easily understood rationale is at the base. Hazards to life, economic gain, nuisance abatement, recreational use, tradition, and aesthetic values must all be considered, but when one factor appears to counter strongly any other, total agreement becomes impossible. Limited justifications have become the rule; this is perhaps the general conclusion one may draw.

There is now no phase of pest control that is not influenced by public agencies. Presumably, the concept of general advantage is accepted as the basis for official decisions, but too many public officials appear not to apply the test of general advantage to the judgments that face them. No other assumption explains the massive mistakes that have been made in the name of responsible pest control.

In prospect for the future are greater conflicts of interest and mounting failure to agree. This is true because pest control operations are becoming both more intensive and more extensive. Many of the operational techniques in use now or probable in the near future are neither precise nor totally controllable. Increasingly significant biological effects will accompany widened pesticide use. It is not at all clear that many pest control officials fully appreciate the changes now taking place.

5 · THE BASIS OF PESTICIDE LEGISLATION

With few exceptions, limits, whether biological or legislative, to the use of pesticide chemicals are not based solely on chemical hazards. Such a judgment is partial, and is in any event a terminal consideration. Good pest control does not and should not begin with a toxic chemical and work backward to the pest. In practice, borne out by the history of pest control procedures, nonchemical factors are far more important than chemical in preventing or alleviating damage caused by animal or plant pests. Current focus on a few major problems of pesticidal use and attempts to correct apparent injustices through legislation are only symptomatic treatments and do not alter the far more serious underlying difficulties, such as the social, economic, and biological limitations. Pesticide legislation can be directed, restricted if need be, more easily than other limitations on pesticide use. Redefinition with broader perspective is an urgent need in pesticide legislation.

The Protection of Person and Property

Hazards to the health of man, his crops, and his domestic animals are fundamental considerations in pest control. There are three basic divisions into which such hazards can be divided—physical injury, disease, and food contamination.

The likelihood of direct injury from predatory animals is now small. Possibly it was never great, but the fear of direct injury formed the first basis for bounty laws, which have been in operation since colonial times. These laws were very early extended to protect livestock and game, and this concern has continued to the present. There have been many hundreds of bounty arrangements, most of them arising from and administered by county or state governments. in essence these laws and ordinances, through cash payments, encouraged individuals to reduce the numbers of a predator species. Shooting and trapping were favored means of capture, but poisoned baits were occasionally used with some success. Most attention has been directed against mountain lions and various members of the dog family—wolves, coyotes, and foxes. Bounty laws directed against these animals are still in force in many states. In later years, coverage expanded to include lesser mammalian predators and some birds. Eagles and magpies are examples. These later extensions were concerned chiefly with game protection. Many thousands of animals are annually submitted for bountying. The search for animals is frequently profitable. A ten- or twelve-dollar payment for each coyote killed, for example, can support an industrious hunter in some plains and mountain states. One man of my acquaintance makes several thousand dollars a year by coyote hunting from an airplane.

Bounties are predicated on the belief that they succeed in lowering the numbers of offending species. Repeated studies have shown that this is not true (Allen, 1954), and, at least where applied for game protection, the bounty system is slowly being discarded. In sheep-raising areas the system will no doubt continue, ineffective and abused as it is commonly believed to be.

Predator control by the federal government began in earnest in 1906, at the request of western livestock growers, and parallels the assumption of greater federal responsibility for the management of public lands, particularly in the western states, where national forests are so important for summer grazing. Often an additional justification has been game protection, and less often, and locally, the elimination of diseases transmissible to livestock.

Responsibility currently rests in the Predator and Rodent Control Division of the Bureau of Sport Fisheries and Wildlife. Many state and county governments have similar programs. The administration is normally in an agricultural department, although funds or direction may also come from a conservation or public-health department. The favored intention at this time is area-wide reductional control, using poisoned baits. There is no question that the system is effective. Faults in the system are discussed elsewhere in this book (see Chapter 16).

Control of field rodents is now assumed by the same agencies responsible for predator removal. Injury to man or animals is, of course, not the justification, although on occasion disease transmission is an important rationale. Damage to agricultural crops or to regenerating forests is the major reason advanced for large-scale campaigns. As in predator control, poisoned baits are used, but a good deal of individual responsibility remains, particularly at local levels. Control officials function largely as advisers on methods and as suppliers of poisoned baits. In recent years emphasis on forest regeneration has resulted in large-scale rodent control efforts on public lands. The ease of distribution by aircraft makes such effort feasible. Large-scale efforts will doubtless remain under governmental jurisdiction, even though private lands may be under treatment. The hazards inherent in the baiting of mammals dictate close supervision and control.

Vector control programs are second only to crop-insect control programs in their variety and scope. The concern lies almost totally with the animal carriers of disease or parasitic organisms. The control of nuisance organisms such as sand flies and gnats is often assumed by the same agencies responsible for disease control. Reduction in numbers of the carrier species is the goal. In a few instances total elimination of a vector has been possible. Vector control specialists are distinguished by their intense efforts to learn a good deal about the biology of target species. It is an axiom in vector control that full knowledge of life cycles will identify the points at which control operations will be most successful. Strong emphasis on habitat manipulation (e.g., ditching of swamps to create flowing water) sets apart this kind of control operation from most others. Recently, the success of broadcast insecticides has dulled the traditional incisiveness of the vector control operator, but he, more than other control specialists, practices his biological trade with awareness of the ramifications of his operations.

The vector control specialist is supported by majority interest, reflected in enabling legislation from the level of city health officials to the World Health Organization. The U.S. Department of Health, Education, and Welfare and the equivalent organization in each state have broad powers to act in the public interest on matters of health. Rarely are their purposes challenged; doubts concerning their methods are less rare and in fact are becoming more common with the wide use of insecticides. In this country, the major problem is unintentional loss of domestic animals or wildlife or adverse alteration of the habitat of desirable animals.

The hazards that a control program may be permitted to create are proportional to public need. In many parts of the United States disease vectors are under as much control as the social structure will permit; only nuisance problems remain to be corrected. In large areas of the earth, however, heroic measures are called for, and used. Nuisances do not compare with the debility and death caused by malaria or bilharzia. The alleviation or correction of a disease or parasite infestation may demand strong measures whose side-effects would not by themselves be tolerated in less-affected countries. In short, vector control measures are strongly conditioned by the attitudes and needs of local human populations. They cannot be judged uniformly throughout the world, as is the tendency in crop-insect control, for example.

I have been discussing thus far in this section the removal of offending species presumed to cause direct injury to person or property. I have implied that there may be some unfortunate effects from attempts to control pests or disease vectors. Legislation aimed at limiting these undesirable side-effects is as varied as the effects. There is no single theme to contain them. However, there is one kind of by-product that has drawn special attention—the effects of the control agents on human health itself.

Chemicals, particularly insecticides, are now used in a manner that does not exclude small amounts of them or their degradation products in foodstuffs. In special instances, carelessness in application or deliberate violation of recommendations results in hazardous amounts of pesticides in human or animal foods. Federal responsibility for the purity of foods rests with the Food and Drug Administration (1958).

Until this century, "purity" of foods was determined by basic sanitation and quality deterioration. Substances used in producing or storing a commodity were either short-lived or restricted to a few products. General contamination has been the rule for only the last decade or two.

The need to restrict residual amounts of pesticides in foods to safe levels has been recognized for a long time. England initiated legal safe levels in 1900. California enacted legislation over forty years ago restricting the levels of arsenical insecticides in foods. Presently, the pattern of "setting a tolerance," a maximum permissible contamination level, stems from the Federal Food, Drug, and Cosmetic Act of 1938. The act prohibits unnecessary additions of poisons, but, where additions are necessary, it authorizes the establishment of tolerable and presumably harmless levels. It remained for the discoveries of World War II and the subsequent surge in chemical pest control to give significance to the legislation.

Immediately following the war it became clear to those involved in pesticide use that the change in kinds of chemicals and, more particularly, the extent of contamination would require new kinds of information. Scientists, advisers, and administrators in industry and government set out to define what the information was to be. The fact of contamination and concern for its effect shortly reached the general public, and in 1950 technical knowledge joined public concern in bringing about an important congressional investigation. The House Select Committee to investigate chemicals in foods was chaired by Congressman Delaney, and the resulting three volumes, containing many testimonials and much new data, are unofficially termed the Delaney reports (1951 and 1952). These reports formed the basis of the Pesticide Amendment to the Federal Food, Drug, and Cosmetic Act, enacted in 1954. This amendment is informally identified with Congressman Miller, a member of the original Select Committee. The "Miller Bill," in effect, defined orderly procedures for setting tolerances and provided penalties for violations. The bill was put into effect in 1955, but new tolerances are still being set and revisions of old tolerances still appear. The Food Additives Amendment of 1958 was a second derivative of the Delaney Committee's investigations.

The Miller Bill applies only to foodstuffs in interstate commerce. Most states now have similar legislation, closely paralleling national laws, to control excessive residues. California, for example, has a "spray residues" law that regulates not only tolerances but also licensing and chemical applications. The many phases of pesticide uses are administratively more closely linked at the state level than at the federal level.

Canadian residue legislation strongly parallels our own. Latin-American governments have only recently begun to standardize their residues laws, which are not yet effective. The World Health Organization has drafted a pesticides code for the benefit of its member nations and is strongly urging the adoption of uniform legislation throughout the world. In Europe, England and West Germany currently maintain the strictest residues legislation.

Governmental specialists have not been able to provide all information necessary for proper administration of pesticide laws. Indeed, in the last decade some responsibilities and reference sources have shifted away from government. Present laws place the burden of proof of pesticide safety on the sponsor of the chemical. Provision of toxicological and performance data has become a major task, performed largely by research specialists in private and university laboratories and by the agricultural experiment stations. The "expert" has become an essential part of the regulatory process.

The National Research Council has become the coordinating body for bringing together the most competent experts to evaluate technical data and to make recommendations in the public interest. Necessarily, the chief preoccupations of the council's committees are controversial or otherwise difficult questions. The body of experts most pertinent to our immediate subject is the Food Protection Committee of the Food and Nutrition Board. Subcommittees provide further refinement of expert coverage. Several reports have been prepared, but one is most germane. This is *Safe Use of Pesticides in Food Production* (Natl. Research Council, 1956). It provides an excellent account of testing and analytical methods and subsequent registration procedures. The report has been widely circulated among the food producing, processing, and distributing industries.

States too call upon experts in decision-making. It has become regular practice for administrators and legislators to call upon expert testimony to assist in judicious formulation and administration of law. An example is the Interim Committee on Agriculture of the California Assembly (Geddes, 1961). Appearance before such committees is routine among specialists, and, without question, is an essential part of technological legislation.

Less routine is the reaction of a state's lawmakers to problems inciting particular controversy. Two "Governor's Committees"—one in Wisconsin (Elvehjem, 1960), the other in California (Mrak, 1960)—resulted from public concern for mounting pesticide problems. The Wisconsin report is organized to include the total impact of pesticide uses. The California committee was more narrowly charged—to investigate the likelihood of hazard from pesticide residues in foods—but in fact its report includes summary statements and recommendations that go beyond this limited purpose. Public response to the occurrence of several widely publicized fish-kills (see, for example, *San Francisco Examiner,* August 1, 1963) prompted Governor Edmund Brown to establish a *permanent* committee on pest control methods and pesticidal effects. The committee, under the direction of the State Administrator of Natural Resources, is not only permanent, but has the widest representation and broadest definition of purpose of any group yet formed to study environmental pollution.

Countermeasures to Dispersal

The normal movements of plant stocks or commodities in commerce, of people in business or pleasure, of traffic in ships or airplanes, and the reduction of natural barriers through altered land use make it inevitable that pest species will be transferred from one area to another. To these transporting "services" must be added the "natural" dispersing powers of organisms, which further increase the likelihood that any area will acquire new pest species. Compound these causes with the disruptions of global war and the agricultural "colonization" of new areas throughout the world, and it becomes evident why new pests appear and establish themselves.

The early colonists in North America were notably free of pestilence. Contact with only native species was short-lived, however. The Hessian fly entered this country in the bedding straw of Hessian mercenaries during the Revolution. The human commensals—lice, fleas, and bedbugs—were imported no one can say quite when. It was not, however, until the latter half of the nineteenth century, following the Civil War and the development of export markets, that alien insects began to appear in numbers. Almost a hundred species of subsequently important insects were introduced into the United States during this period. A few familiar examples are the Argentine ant, horn fly, boll weevil, European red mite, gypsy moth, and San Jose scale.

Near-catastrophe in California citrus orchards because of San Jose scale gave rise in 1881 to the first quarantine law in America (Camp, 1956). In 1889 the Massachusetts legislature made the first appropriation for the control and eradication of an immigrant pest—the gypsy moth. Neither quarantine action was by itself effective. San Jose scale was controlled in a short time by a biological control agent; the gypsy moth at this time occupies several states and at best can be considered only "contained."

Quarantine means exclusion. It may be applied to prevent both initial or recurring ingress. Following establishment of an introduced species, it may take the form either of containment—to prevent further spread—or of eradication—total elimination of the species. The three phases are not equally successful. Under current conditions we might say that exclusion is effective in 90 per cent of incipient establishments; containment, in perhaps ten per cent of cases after successful establishment; and eradication, in perhaps one per cent or less.

The first federal law to deal with quarantine action was the Insect Pest Act of 1905. It prohibited importation or movement of insect species known to be hazardous to crop plants. The act is particularly meaningful but also difficult to enforce in this era of aircraft travel. However, the Insect Pest Act was relatively limited in scope. The Plant Quarantine Act of 1912 provided the Secretary of Agriculture with broad powers to enforce such regulations as he deemed necessary to protect the agricultural economy from foreign importations of insects and disease organisms. Important points within these broad powers are inspection at points of entry, inspection without warrant while in transit, strong penalties for violations, and seizure and destruction. The act has since been amended several times. Related acts deal chiefly with inspection procedures at international borders and the prevention of export of organisms potentially damaging to the agriculture of other countries. Since 1936, postal laws have been especially effective in supplementing quarantine laws (Stefferud, 1952).

Inspection procedures are remarkably successful considering the great amount of commercial traffic. Some 3000 species of insects have been found in aircraft. Interceptions of improper plant materials increased from 137,000 to 291,000 in the decade 1948–57. Since establishment of the Plant Quarantine Act, about thirty economically important species of insects have become established, and of these probably not over four or five can be credited to the last fifteen years.

Once an establishment is discovered, the decision must be made whether to contain or to eradicate. Ordinarily, public hearings are held to discover the likelihood of spread and damage, and an assessment of probable hazard is followed by a quarantine order. Since initial establishment is apt to be within single states or regions, the federal control operation must necessarily rely heavily on state action. Most states have legislation enabling the same inspection, seizure, and control operations exercised by federal officers. In the last decade, through the cooperation of state and federal departments of agriculture, joint ventures have been greatly facilitated (Popham and Hall, 1958). It is now no longer accurate to speak meaningfully of separate responsibilities of state and federal quarantine agencies. They are tied by cost-sharing interlocking operations and, most important although largely implied, by an awareness that one cannot function adequately without the other. These close ties are well illustrated in the recent containment operations against the gypsy moth in the Middle Atlantic states and in the Japanese beetle and Dutch elm disease control campaigns in the North Central states. They existed also in the only successful eradication attempts of the last two decades—Hall scale in California, Khapra beetle in California and the Southwest, Mediterranean fruit fly in Florida, and screw-worm in the Southeast. The cooperation initially displayed in the attempted eradication of the imported fire ant in the South may be benevolently described as overenthusiastic.

A few states independently maintain rigid inspection procedures. California is the most stringent; in some instances strict quarantine is maintained on county levels (Lemmon, 1957). Every auto traveler entering the state is stopped at border entry points and questioned about the plant material he is carrying. Freight and postal shipments of seeds and nursery stocks must often be routed through county agricultural commissioners for certification.

Natural dispersal and changes in land use, combined with the complications of human travel and commerce, dictate against national quarantine and control measures at international land borders. In general, there are two methods by which international measures are effected. One is an area-wide or regional approach, consisting normally of contiguous national governments combating a common pest species. The Anti-Locust Commission, comprising both European and North African nations, is such an organization. Bolivia, Argentina, Brazil, and

Chile have a comparable organization. Tsetse fly control is organized in a similar way in Central Africa. The second method may take the form of exclusion rather than control, as with the citrus blackfly. This insect has thus far not gained entry into the United States, although it is common in Mexico. The closest kind of alliance between the two governments has been necessary to limit its dispersal.

Although an international control organization may be limited geographically, it may not limit itself to single pest species. Such is the case with the European and Mediterranean Plant Protection Organization. This group is an outgrowth of a more limited one—the International Committee for the Control of the Colorado Beetle. Such organizations arrange for research on problem species, the exchange of information and personnel, the provision of operating funds, and the execution of control programs.

The only truly global organization does not provide all these services. The Food and Agricultural Organization of the United Nations has provided a world reporting service since 1951 (*The Plant Protection Bulletin*) and has formulated uniform regulations and quarantine procedures whose adoption is urged upon member nations. These purposes alone, if fully realized, justify the activities of F.A.O. It is probable that this comparatively new emphasis will serve as a basis for yet closer ties among nations in the control of pests.

In a less direct way the World Health Organization also contributes to quarantine and control legislation. The many vector control programs throughout the world, aided in some way by W.H.O., operate on the premises of reduction in numbers of vector species and alterations in human habitation in order to minimize contact between the disease-carrying vectors and human beings. Where operations succeed in containment or eradication, locally or regionally, they become "quarantine" actions. The best example is the total removal of a fairly well established population in Brazil of the yellow fever mosquito (*Anopheles gambiae*), presumably brought from French West Africa.

Protection of Food and Fiber Supply

The control of pests affecting commodity production, storage, and distribution was traditionally the responsibility of the individual. Early legislation concerned invasion and establishment of pest species, and only incidentally affected local control problems. The Insect Pest Act of 1905 recognized the value of inhibiting traffic of insects between crop-producing areas. The Plant Quarantine Act of 1912 implemented and broadened the prior law. The exclusion clause in these acts was enforced by inspection, seizure, and condemnation procedures to ensure reduced traffic in insect species. Containment and eradication programs are simply extensions of the exclusion idea.

Many years after the passage of the Plant Quarantine Act, Congress recognized that general pest control was in the public interest. The basic law approved by Congress in 1937 delegated the responsibility for control to the Secretary of Agriculture in these words: "The Secretary of Agriculture, in cooperation with authorities of the states concerned, organizations or individuals, is authorized and directed to apply such methods for the control of incipient or emergency outbreaks of insect pests or plant diseases . . . as may be necessary." The key words are *cooperation* and *directed*. Together they explain the present manner of operation of the federal control agencies. A later law (1944) allowed the Secretary of Agriculture to act *independently* as deemed necessary in the control of any infestation. In 1947 this sequence of authorization and operation was applied to forest insects. The Secretary of Agriculture, in language very similar to the above quotation, was authorized to conduct surveys, to investigate control procedures, and to carry out such measures as necessary to accomplish control objectives. The value of forests to the public thereby received the same recognition as that of croplands (Camp, 1956).

The large-scale control programs (gypsy moth, fire ant, spruce budworm, and grasshopper) are based on the foregoing directives to the Secretary of Agriculture. The Plant Pest Control Division of the U.S. Department of Agriculture is responsible for carrying out decisions to control on croplands. The Forest Service assumes the parallel responsibility on forested lands.

Interagency committees are maintained at both policy and operating levels when control programs are apt to affect other interests. Most important of these are health (U.S. Department of Health, Education, and Welfare) and wildlife loss (U.S. Fish and Wildlife Service).

To this point, legislation has been identified only with the need to restrict movements of pests and to suppress populations of offending species. Both purposes are legislatively defined as matters of public interest. Radical changes in the character of and ability to disperse pesticides have brought a need for protection against the control agents themselves—again defined in the public interest. Two legislative approaches have been taken: one, to define the manner of use of hazardous pesticides; the other, to deal with residual amounts of such chemicals in foodstuffs. Human health has been the primary consideration in both approaches. Restrictions on amounts of residues under the Miller Bill and Food Additives Amendment have already been discussed. Restrictions on initial use are the responsibility of the Plant Pest Control Division of the Department of Agriculture (U.S. Dept. Agr., 1948——). The current reference law is the Federal Insecticide, Fungicide, and Rodenticide Act of 1947. In 1959 this law was amended to include other classes of agricultural chemicals—nematocides, plant regulators, defoliants, and desiccants.

The total process of controlling pesticide uses is complex and ranges well beyond the federal and state agencies concerned. In fact, it is reasonably accurate to state that these agencies do not directly control use. They do control marketing, however. They do so primarily by registration of chemicals. Registration is preceded by an accumulation of data on both hazard and use. Label statements on packaged chemicals must accord with the best information available. Appropriate warnings must appear on each label. These statements must be clear and explicit, based on determinations that use of the product according to instructions is safe to the user and to the public and will not result in unlawful residues. Most states now have agencies that register chemicals in the same manner. California, for example, has registered over 12,000 pesticidal products (Lemmon, 1957). At the state level as well as the federal, "registration" laws are closely linked to "residue" laws. Compliance with law is tested by periodic sampling of marketed chemicals to ensure that label guarantees are met.

Recent federal and state registration laws have the effect of requiring the chemical manufacturer or formulator to prove safety before use. Before these laws were established it was up to the agencies to prove hazard. The reversal of responsibility has resulted in an important limitation on use that only secondarily can be called a legislative restriction. Registration requirements that are fully met are costly. When coupled with the costs of development and field testing, costs can become prohibitive. The small investor cannot play the role of entrepreneur.

There are other kinds of pressures which force legislative action on pesticides. In general they are public attitude, war requirements, market conditions, and so on. Public outlook is particularly important, but it is difficult to analyze except in broad terms. It lacks—according to the views of some specialists—the perspective that full knowledge might bring. A. C. Worrell, of Yale University, has provided a rare attempt to integrate the diverse elements involved in decisions to use pesticide chemicals (Worrell, 1960). His paper is the best of its kind I have seen. The book by Gunther and Jeppson (1960) accomplishes much the same purpose, but represents too strongly the views of the pesticide industry. Rachel Carson's *Silent Spring* (1962) has been the only account of pesticidal hazards that has been successfully directed to the general public.

The Protection of Desirable Forms

Relatively few kinds of animals and plants are inimical to human welfare. A very great deal of effort, however, is expended to limit the depredations and misery caused by these few species. The majority of legislation reflects concern about these species. The remaining species —millions—are not favored by legislative concern, except for a few which receive preferential treatment. Normally, among animals, these exceptions are species that provide sport, yield a specialty food item, appeal to curiosity, or are distinguished by their rarity. Plant species may be protected for some of the same reasons, but more often they are protected as shade or decorative trees. Distinctive assemblages of plants and animals may be protected in parks and refuges by public desire, although, more often, particular species and uses are the subject of individual legislation.

The widely divergent purposes of "nature" protection are reflected in law. Legislation is generally inconsistent in its measures to protect wildlife except for the popular game species. And except as regards these species, there is also little strength in legislation. Wildlife protection, in contrast to pest control, depends on public attitude for its continuation.

The U.S. Fish and Wildlife Service is the most important single governmental agency concerned with wildlife protection. The statutes that govern its operation are broadly written, but its efforts are directed largely toward management of the few species of birds, mammals, or fishes popularly rated as game, food, or pests (note the redundancy in the name Fish and Wildlife Service). The state conservation departments parallel it throughout, with two important qualifications. First, game species receive the greatest emphasis; pest control is not a major concern with most such agencies, nor is nature protection in the widest sense. Secondly, although most state codes are written to include broad responsibility, only a few states approach full assumption of this responsibility. Moreover, enforcement of code violations is difficult and frequently unpopular.

With reference to the federal pesticide legislation I have already discussed, one important point must be made. Wildlife values are recognized in laws not administered by the U.S. Fish and Wildlife Service. Tangibly, this is reflected in the labeling requirements of pesticide chemicals under the Federal Insecticide, Fungicide, and Rodenticide Act of 1947. Where clear hazard to wildlife exists, warnings must be included on each label. No penalties are provided for violations of recommendations or for producing unintended damage within the limits of recommended use.

In some states wildlife hazard is given strong consideration in both chemical registration and the procedures recommended. California, Connecticut, and Wisconsin are examples. In these states the full range of hazard is given more weight than the law basically requires. (Agencies performing in such fashion deserve especial recognition for the difficult task they assume.) The U.S. Fish and Wildlife Service cooperates with other agencies in providing information and judgment on wildlife hazard connected with crop-insect and forest-insect control, water-pollution control, and health-related control programs. Relationships with other agencies are sometimes formalized (e.g., the present committee liaison with the Plant Pest Control Division and the Forest Service), but their success is more apt to be based on the competencies and personalities of individuals. The Fish and Wildlife Service does, however, have two formal areas of pesticide responsibility. The older one falls under the Predator and Rodent Control Division, whose function is the control of or alleviation of damage caused by vertebrate animals. The development of new methods of noxious-vertebrate control and the assessment of hazard to wildlife from all pesticide uses are the responsibility of the Branch of Wildlife Research. Supporting funds have enabled a beginning to be made on research investigations on a broad front. The important word is *beginning;* research of this kind is extremely complex—not easily subject to the neat experimental designs of the developmental chemist. Nonetheless, a start has been made and a few patterns are emerging. A much larger program will be required in (I estimate) three or four years.

Interagency relationships are not always comfortable, nor can private groups and public agencies always reach accord. The recent conduct of large-scale "mass" spraying programs has not been distinguished by effective cooperation either among agencies or between agencies and the public. The serious discord following the fire ant control program, after a period of recriminations, resulted in stronger agency ties and more judicious appeals for public understanding. However, it prompted a legislative attempt to enforce both close interagency operation and a broader base for decision-making. The bill in question is H.R. 11502, "to provide for advance consultations with the Fish and Wildlife Service and with State wildlife agencies before the beginning of any federal program involving the use of pesticides or other chemicals designed for mass biological controls." Extensive hearings on the Chemical Pesticides Coordination Act were held in May, 1960, but no further action was taken by the 86th Congress. Essentially the same measure was put before Congress shortly after convening in 1963, but action was postponed until after the appearance of the President's Science Advisory Committee report on use of pesticides (Kennedy, 1963). Strong public pressure remains for enforced cooperation among public agencies. The practicality of such a measure if adopted remains to be seen. Its most probable effect will be to slow the "enthusiasm" of advocates of eradication programs. For less extreme positions, it will probably accomplish little that cannot be accomplished as well by other, more pleasant means.

Congressional pressure for the Pesticides Coordination Act was dropped temporarily following the formation of the Federal Pest Control Review Board in October, 1961. This organization "... will insure that the framework of the objectives of the [pesticide control] programs are [sic] sound and that the most effective, economical and safest procedures are followed." The purposes of this new board are those envisioned in the Pesticides Coordination Act. It remains to be seen whether its purposes will be satisfactorily accomplished. To date, no clear evidence of its value has been forthcoming.

Private actions sometimes influence pesticide operations and legislation a great deal. Of course, in the long run all legislation is a reflection of private judgments, but in a more immediate way individuals and groups determine the boundaries of pesticide disputes and subsequent legislation. I have already described a case in point, the citizens' suit against the gypsy moth control program (see Chapter 3). Citizens' groups have in a number of instances been organized and have effectively argued for wildlife values in several states. Private organizations such as foundations and scientific societies have lent expert counsel and authority for matters under their aegis. Their effect on policy and legislation is indirect but real (e.g., Conserv. Foundation, 1960).

One of the most influential private bodies is the National Research Council. It is not a governmental body, as it might appear to be, but it is an adviser to government. The National Research Council recently turned its attention to pesticide-wildlife relationships, and now has a functioning Committee on Pest Control and Wildlife Relationships. The summaries and recommendations as they appeared might be expected to have great weight in determining the pattern of legislation in the pesticide field. Such an expectation has not been met in two of the three reports that have thus far appeared (Natl. Research Council, 1962). Although coming from a distinguished committee, the reports are so general as to be meaningless, are unscholarly, and, most curiously, evade the central reason for the committee's formation (the clear evaluation of wildlife-pesticide relationships). Hopefully, the third and last part (research needs), to appear in 1963 [after completion of this manuscript], may more nearly satisfy its purposes.

The functional counterpart of the National Research Council's Committee on Pest Control and Wildlife Relationships (although with closer ties to government) in Great Britain is the Nature Conservancy. It is highly regarded for its responsible evaluations. At its suggestion, for example, the use in Britain of aldrin, dieldrin, and heptachlor as "seed dressings" was discontinued in 1961 after clear hazard and actual damage from their agricultural use were demonstrated. A suggestion for discontinuance of use of these chemicals would not be acted upon so positively in this country.

III · THE RESPONSES OF ANIMALS
TO CHEMICAL CONTROL

6 · THE INVERTEBRATES—
WATER, SOIL, AND SURFACE

The invertebrate animals comprise by far the largest number and variety of living things. In fact the phylum Arthropoda, which includes the insects, contains about 90 per cent of all animal species. Terrestrial invertebrates are greater in number while aquatic forms include.a greater variety. Most of our pests, parasites, and disease vectors—the targets of pest control—are invertebrate animals. However important these species may be as pests, the majority of invertebrates can be looked upon as beneficial forms—at their simplest, interesting, and, more finely analyzed, both economically desirable and truly necessary in the maintenance of animal food chains. The complexity of the interactions between invertebrate animals and their physical and biological environment is perceived but incompletely understood by scientists. Only in the simplest of ways are the interactions manipulated for the special favor of man.

No section of the earth's surface on which man is permanently resident is free of invertebrate animals, nor in any area is he free of the need for them. The present discussion centers on the invertebrate animals living in water or in or on the soil. Summarizing the highly intricate relationships within a few pages of text is presumptuous, but an attempt must be made.

Aquatic Invertebrates

There are three aquatic environments of special interest to us here—freshwater streams and ponds, marginal freshwater and brackish marshes, and the shallow coastal margins of the sea. Knowledge of the effects of pest control on these habitats declines in the order named. A good portion of this knowledge derives from planned insecticidal pollution whose hazards were foreseen. This is the case in blackfly, mosquito, and sand fly control. An increasing portion stems from studies on unintended contaminations associated with terrestrial-insect control. Hazards from this source have not always been foreseen, and unwanted mortality has often been spectacular.

A number of studies in the middle and late 1940's centered on the hazards of DDT in widespread use over stream and pond environments. Notable are those under the sponsorship of the Public Health Service and the cooperative programs of the (then) Bureau of Entomology and Plant Quarantine and the Fish and Wildlife Service. The papers of E. L. Bishop (1949), C. H. Hoffmann (Hoffmann *et al.*, 1949; Hoffmann and Surber, 1949a, 1949b), C. M. Tarzwell (1950), and P. R. Springer (Springer and Webster, 1949, among others) are particularly important and there are many others. The realization that DDT possessed wide-spectrum toxicity stimulated many field investigations of its effects on nontarget species. The body of data from this period of intensive study with DDT remains the best available. Later studies have amplified our knowledge of DDT effects, but, at least as far as direct toxicity is concerned, they have not altered the earlier major conclusions. The same attention has not been given to other chemicals and means of application. In part this inattention seems justified because of the thoroughness of the DDT work. But it is unjustified when one considers the magnitude of the indirect or delayed effects not foreseen in the original studies. The illusion of safety that resulted from the studies persisted in recommending agencies for several years. Only recently have public repercussions stirred this lethargy.

From these studies the following general statements can be made, which describe reasonably well the nature of biological effects from DDT applications over freshwater streams and brackish marshes.

Most intentional applications of DDT to water surfaces have been made at 0.2 to 0.5 pound per acre. Before continued exposure produced resistance and other complications, these levels succeeded in controlling target organisms (i.e., producing drastic reductions temporarily).

Unintentional applications to water result from treatments on land near water borders at higher doses; 0.5 to 1.5 pounds per acre would be the normal range of application. These applications too serve their intended purpose reasonably well. Because in most cases chemical applications are not made directly to water, it is probable that the amounts reaching water surfaces through runoff are only a small fraction of the amounts applied to the land—perhaps one-fifth or less. Direct accidental application at full doses from poor pilotage would constitute an exception.

Critical influences in toxicity are the volume and rate of water flow. Shallow impounded waters yield most severe reductions. Strong flushing and dilution resulting from tidal or stream flows lessen the immediate impact and duration of effect.

Unintended effects can result from much less poison than is commonly used or than is necessary to control a pest. The effects at these lesser amounts are not equal. Susceptibility depends on the animal group, its life habits, and the manner of exposure.

In general, arthropods are sensitive to small amounts of chemical contamination. Particularly vulnerable are insect larvae, marsh spiders, and all crustaceans. Surface insects (e.g., dytiscids, corixids) are slightly less sensitive, but nonetheless are drastically affected by rates above 0.5 pound per acre. Certain kinds of worms (for example, oligochaetes) are about as responsive as the surface insects. Planktonic forms range in susceptibility according to group; small crustaceans are the most sensitive (note, for example, the extremely high *Daphnia* sensitivities recorded in Table 10). Adult mollusks—mussels and snails —seem highly resistant.

Rates of population recovery vary with the extent of reduction (a function of application rate), the faunal complex, the frequency of repeated applications, and the physical character of the aquatic environment. Recovery in most groups exposed to applications under 0.5 pound per acre seems to be sufficiently rapid to discount such loss. Above this amount, euphemistic appraisals that recovery rates are rapid enough to repopulate in reasonable periods do not seem supported. Even at 0.5 pound per acre in some situations (as I will shortly describe), optimistic appraisals seem questionable.

TABLE 10

Estimated concentrations of organic insecticides required to immobilize *Daphnia magna* in 50 hours (from Anderson, 1960)

Pesticide	Concentration in water (parts per billion)
EPN	0.1
Parathion	0.8
Malathion	0.9
DDT	1.4
Methoxychlor	3.6
Aldrin	29.2
Heptachlor	57.7
Dieldrin	330.0

I would judge, with the benefit of hindsight, that large-scale DDT applications contribute only a delaying action against the pest species. Intended results in control have been achieved largely in those schemes where repeated exposure was not required. Where repeated, the problems of residues and insecticide resistance have dictated against continuing use of DDT. The limited usefulness of DDT repetitions is further enforced by unintended effects on other species of animals. The effects on aquatic organisms have been more pronounced than at first believed, and recovery after treatment seems much slower than anticipated. The extension of treatment to millions of acres rather than the few used in experimentation or in local pest control has been accompanied by unwanted effects equal in magnitude to the damage caused by pest species.

The complications of repetitive vector control have already forced substitution of other chemicals or a return to environmental controls in this country. This conclusion does not yet apply on a global scale, but will in time. Forest-insect control does not normally demand repeated chemical treatment at frequent intervals. Problems of residue and resistance have thus been avoided there. Other biotic problems remain, and in fact are becoming magnified.

Let me illustrate how serious these complications can be with reference to aquatic organisms in major streams coursing through treated forest lands. The examples come from salmon streams exposed to treatments for budworm at 0.5 pound of DDT per acre in New Brunswick (Ide, 1956) and at 1 pound per acre in British Columbia (Crouter and Vernon, 1959).

In the earlier study, immediate effects on stream fauna from a DDT application in mid-June were marked. Following immediate kill there was negligible insect emergence for a month after spraying. In late summer there was a sharp increase in numbers, but almost all were small chironomids. Insects emerging in the year of spraying were almost all small, in contrast to the preceding year, when a large portion were large forms. The actual volume of production following spraying was much reduced. Less than one-half the usual bulk of insects was produced in the year following spraying. There were significant differences from normal in the insect types composing the biomass. The same pattern obtained in the second year following spraying, although a slight recovery of some large forms seemed to be taking place. The most conspicuous features of the sprayed streams were the large number of chironomids and the almost complete absence of caddisflies. The same kind of increase in numbers of small flies following stream spraying has been observed in blackfly control campaigns. The dominance of the small flies reduced total productivity and, through selective elimination of the larger forms, denied the larger young salmon satisfactory food materials.

The second study will also be alluded to subsequently in connection with toxicity to fish. It was a particularly good study designed to analyze several deficiencies in application procedures for forest-insect control. Spraying near streams was deliberately avoided. Nonetheless, effects on invertebrate bottom fauna were dramatic, much the same as were observed where less care was taken. Several sites were studied. Contrasts in numbers and weights per square foot of stream-bottom invertebrates are shown in Table 11 for two sites before and after spraying.

Larger forms were drastically reduced. Caddisflies were totally eliminated; stoneflies were half as plentiful five months later; and mayflies were about one-tenth as common. In summary, the major fish-food organisms in affected streams were drastically reduced and were of insignificant food value for the year of spraying. The slow rates of recovery indicated that the same conditions would obtain in the year following spraying.

TABLE 11

The weight and number of organisms per square foot of stream bottom before and after spraying with DDT for budworm in British Columbia (after Crouter and Vernon, 1959). Percentage change calculated from prespray figures.

	Prior to spray	Immediately after	% change	Five months after	% change
			Sprayed areas		
I. Numbers	42.0	0.78	−98	2.2	−95
Weights (grams)	0.0285	0.0038	−87	0.0021	−92
II. Numbers	26.9	5.21	−81	5.9	−78
Weights	0.0226	0.0029	−87	0.0018	−92
			Unsprayed area		
Numbers	24.5	12.6	−49	13.0	−47
Weights	0.0287	0.0212	−26	0.0295	+ 3

The foregoing examples and commentary apply to most stream and marsh environments. Special consideration must be reserved for organisms in salt water—tidal flows and the sea itself. The effects of pesticides in fresh water are measured largely by alterations in insect numbers. There are essentially no insects in salt water. The predominant arthropod class in this environment is the Crustacea, a group of organisms particularly sensitive to chlorinated hydrocarbon insecticides (Tables 12 and 13). Reduction in numbers is catastrophic when crustaceans—crabs, sand fleas, isopods, and so on—are subjected to pesticides in tidal marshes, as well as in all other environments occupied by crustaceans. The other important phyla are somewhat less affected—worms moderately so, and adult mollusks seemingly not at all at low exposures.

More critical here than sensitivity is the likelihood of exposure. It will assuredly continue to be acute in vector control operations on tidal marshes. Both the type of chemical and the manner of control could markedly change the degree of hazard. Vector control specialists seem in fact to be increasingly emphasizing environmental manipulation as the basic control procedure. I believe we may look forward to less hazard to aquatic organisms in vector control procedures.

The margins of the sea produce valuable amounts of mollusks, crustaceans, and fishes. Other groups of plants and animals in the sea are less important economically. Nonetheless many are valuable for particular purposes, varying with region of the earth and cultural tradition. Sea cucumber is a delicacy to Chinese, for example, although most Americans would not find it palatable.

TABLE 12

Approximate concentration of four common pesticides required to immobilize small stone crabs within 24 hours (from Butler, 1962)

Pesticide	Concentration in water (parts per billion)
Endrin	10
Dieldrin	10
DDT	10
Sevin	1,000

TABLE 13

Approximate 48-hour median tolerance limits (TL_m) values for white shrimp —a commercially valuable species—in the presence of some common pesticides (from Butler, 1962)

Pesticide	Concentration in water (parts per billion)
Endrin	6.5
Sevin	13.0
DDT	25–50
Dieldrin	25–50
Toxaphene	75–90
TDE	75–90
Heptachlor	250.0

The question arises whether pest control chemicals extend their influence to the sea. The best present judgment is, "not very much—yet."

Again the likelihood of exposure is critical. Commercial fisheries still depend on removing a portion of a naturally produced population. Management is normally only of the simplest order—regulation of the size and number taken. Only recently has there been any general interest in regulating variety, amounts produced, and quality in the manner that is now common practice among agriculturalists.

The only important exception is the oyster industry, which for some years has controlled spat planting, sought favorable sites for culture, and attempted predatory-animal control. It is only in this industry that direct, intentional contamination of shallow sea water has been seriously studied. The Bureau of Commercial Fisheries now maintains an active program in New England and on the Gulf Coast to study possible means of controlling oyster competitors. Among common pesticides, DDT, lindane, Sevin, and dieldrin have all proved useful in preventing the establishment of barnacles. V. L. Loosanoff, of the Milford (Connecticut) Laboratory of the Bureau of Commercial Fisheries, feels that pesticides can be safely used in oyster culture. However, the field of study is new and little is known of the effects that toxic contamination produces. Effects are certainly present. At 0.05 part per million (p.p.m.), DDT caused 90 per cent larval mortality in oysters and essentially prevented the growth of survivors. Growth rates were impeded at concentrations of 0.025 p.p.m. (Loosanoff, 1960). Lindane has been used safely in the control of crustaceans in oyster beds. Compounds that inhibit toxic bacteria may enhance the growth rate of clams and oysters. However, marine phytoplankton, a staple food source for bivalve mollusks, may be harmed by many toxic chemicals (Table 14), some of which kill when present in only very small amounts (Butler, 1962).

TABLE 14

The highest concentration of herbicides and insecticides tolerated by five kinds of phytoplankton used by molluscan larvae as food (from Butler, 1962)

Pesticide	Concentration in water (parts per billion)
Herbicides	
Monuron	0.02
Diuron	0.04
Lignasan	0.06
Neburon	0.40
Fenuron	290.00
Insecticides	
Sevin	100.0
Lindane	500.0
DDT	1,000.0
Dipterex	10,000.0

The main emphasis in chemical uses in the sea has been to eliminate competing organisms from the surfaces of economically valuable animals. Control of predators, regulation of growth rates, and so on, cannot yet be widely practiced. There are surfaces other than those of shellfish that we wish to keep entirely free of marine organisms, namely ships' bottoms and wharf pilings. Perhaps these purposes are not properly included here. I need only comment that mollusks and crustaceans are the chief offending groups and that research on antifouling compounds has been fruitful. Normally employed are toxic "paints" that slowly release slight amounts of copper. All such treatments act at or very near the treated surface and cannot be considered general contaminants.

Potentially more important is unintended contamination of shallow coastal waters. The richness of life at the sea's margin derives in large part from nutrients carried into it by freshwater streams. If the stream is very large, its fertilizing influence may be felt many miles from land. When the stream has coursed through many miles of agricultural lands, it is quite likely to acquire measurable amounts of pesticides and other contaminants. The Public Health Service has clearly established that small amounts of pesticide chemicals exist in waters at the mouths of some of the most important rivers in this country. These amounts, although small, are increasing; so also is the variety of chemicals found (see Chapter 13). That such chemicals exist at all in major drainages is one problem. Their effect, given their presence, is another. It would seem that the amounts thus far determined are too small to produce toxic effects directly. Nonetheless, there remains the strong likelihood that certain of the most susceptible invertebrates can be affected. The shrimp fisheries off coastal Louisiana are one situation where the question of possible pesticidal effects has been raised (Gresham, 1962). Pesticidal effects have also been suggested as an explanation for some die-offs among inshore crab and mollusk populations. I know of no convincing data to support these allegations, but I do believe that such loss is not unlikely. The Gulf Coast would seem to offer the essential requisites: a major drainage flowing through agricultural land to the sea, and shallow, productive, reasonably stable marine waters. At present the amounts of pesticides that have been found are not considered hazardous to human beings, nor to fish under most circumstances. Measured amounts within the range of crustacean susceptibility have been found on occasion, but no biological studies have been made to confirm effects at these levels. Based on the assumption that pesticidal residues will continue to be present (and perhaps rise) in major river outlets, we may foresee wide-scale effects for two reasons: There is no escape from them in aquatic environments; food-chain organisms pick up, store, and accumulate stable pesticides found in their food medium, thus becoming toxic to the animals that feed on them. These reasons would seem to offer more serious concern than heavy mortality from direct occasional contamination.

Soil-inhabiting Invertebrates

If soil were regarded simply as an inert substance that will prop up plants and hold fertilizer, then control of soil pests should be directed toward inhibiting the numbers of those few organisms causing important reductions in yield or quality. By extension, the soil would be considered merely a convenient medium on which to produce foods and fibers until such time as we can divorce ourselves from it. We would depend on it because we presently have no efficient alternative. Production in these terms is a form of mining, only slightly more complicated than withdrawing metals from the earth. This view is supported in practice by the majority of fertilization and pest control practices. Such an attitude ignores the basic contribution of living soil—the decomposition of organic matter. The return of once-living things to the soil improves texture and water retention and, most important, again makes available in usable form the simpler chemicals necessary to continue life.

In elementary biology we describe the carbon and nitrogen cycles and discuss trace-mineral requirements without identifying them directly with practical production problems. By both student and agriculturalist the unavoidable truths of the organic cycles seem not to be appreciated. Richness of soil comes from the life within it. Productive soil is teeming with life. When an agriculturalist uses crop yields to assess soil productivity, he is reporting an indirect measure of the amount and variety of soil organisms present within it. A relatively sterile soil produces a relatively poor crop. The transfer of fertilizers and mineral nutrients from one point on earth to another has the effect of increasing soil fertility in the favored site but in no way changes the basic premise. "Advanced" cropping systems, which have as their goal the selective utilization of small fractions of the total biological productivity and which attempt to enlarge only this fraction, can be self-defeating, especially if they follow the path of "clean" cultivation—burning, removing the surface organic matter, or otherwise contributing to a decline in humus content with its attendant plant and animal life. Under these conditions decline in soil fertility can be predicted. The decline is hastened when toxic chemicals are added to the soil—chemicals that interfere with decomposition processes or create imbalances in the soil complex. Frequently, new problems arise where none or small ones previously existed. I strongly suspect that soil biology is the single most neglected area of pest control research. At the conclusion of this section I will list several criticisms of current philosophies and techniques of soil-pest control that give the reasons for this suspicion.

The numbers of organisms in the soil are staggering. The biomass of soil organisms, taken collectively, constitutes well over half of all living matter. Decomposing organisms (bacteria, fungi) comprise the greater number, but detritus feeders, parasites, and predators probably constitute a larger mass. The species and variety of larger organisms tend to differ according to soil, crop type, and region of the earth. There is a greater uniformity in types of microbial organisms. If we consider energy requirements rather than numbers, kinds, or amounts, we can derive the "fuel" requirements of entire complexes. Metabolic activity is the best available measure of productivity and degradation rate (MacFadyen, 1957). In soils where the respective groups are at normal abundance, an approximate order of importance according to metabolic activity would be:

Bacteria	Mollusks
Nematodes *	Mites
Fungi *	Fly larvae *
Protozoans	Earthworms
Collembolans	Spiders
Beetles *	Harvestmen

It should be noted here that the above order is approximate, and, further, that variation occurs from season to season and region to region according to soil type. Moreover, it excludes vertebrates that burrow or find shelter in soil.

The commonest soil pests belong in the groups marked with an asterisk in the above list. In general, damage results within the soil from the prevention of seed germination (primarily by fungi) and from physical injury to root systems (sometimes producing secondary injury through disease transmission). Less commonly, there is no within-soil damage but control is directed toward killing early stages in the life cycles of insects that would later produce above-surface injury. Microbial organisms and nematodes are confined to the soil throughout their life histories; insects normally live in the soil only during their larval stages. Most effort at control of soil animals has been directed toward insect larvae that damage root systems. The magnitude of nematode damage has come to be appreciated only in very recent years, and greatly expanded research in nematology has followed.

Specific targets of control are chiefly larvae of the insect orders Lepidoptera (butterflies and moths), Coleoptera (beetles), and Diptera (flies). The three best-known soil pests are cutworms (moth larvae), wireworms (tenebrionid and elaterid beetle larvae), and white grubs (Japanese beetle, June beetles, and others). As with the nematodes, the type of cultivation normally affects the numbers. Changing the crops or cultivation practice is the best method of reducing numbers of all these forms. Where changes cannot be made, chemical control is practiced. Chemicals may be applied directly to soil, where they are generally distributed by plowing and disking throughout the top few inches of soil. Volatile chemicals may be drilled into soil or applied in spots or rows, from which their effects spread. In some instances protective strips of nonvolatile chemicals are placed in furrows paralleling rows of plants. Chemicals can also be applied directly to seed. Fungicides have long been used in this manner. Now it is common practice to add insecticides as well to the slurry in which seeds are mixed. Several chlorinated hydrocarbon insecticides have been successfully used. The best known is BHC or its gamma isomer, lindane, for wireworm control. The sphere of action of seed-treatment chemicals is limited; hence, their effect on soil microorganisms is not general.

The best evidence that mortality occurs as a result of soil-pest control is the reduction in numbers of the target species, but the customary measure is the crop yields that follow control applications. The majority of assessments of soil-pesticide efficacy have been made by these two measures of control success. Only rarely and recently have nontarget numbers been studied as well. Lilly (1956) points out how ecological study has been generally neglected in recent years. Nonetheless, he notes increasing use of soil insecticides by growers (particularly of row crops—corn, sugar cane, etc.) and predicts far greater use.

The most widely used soil insecticides at present are chlorinated hydrocarbons, characterized by low phytotoxicity and long residual action. The most serious hazard to nontarget invertebrates arises from high application rates directly to soil. A variation of this is the slow accumulation of insecticides in soil when application (even though in low amount) is repeated before chemicals from previous applications have been completely degraded. (This point is expanded later in Chapter 13; I need only comment here that almost nothing is known about the ecological consequences of such accumulations.) Some concern is expressed by forest entomologists about the possible effects that forest spraying might have on soil microorganisms. This concern stems from the realization that forest communities must remain complex to permit the most satisfactory pest control.

DDT has been applied to ground surfaces at rates as high as 25 pounds per acre to control white grub. Dieldrin has been used at rates of 5 pounds per acre for white-fringed beetle. Several wide-scale campaigns employ aldrin, dieldrin, or heptachlor at rates up to 3 pounds per acre. Vertebrate mortality has been associated, both directly and indirectly, with all these applications. That there are profound effects on soil organisms I have no doubt, but I know of no serious studies on them.

Surface applications to be worked into the soil are normally at the rate of 0.5 to 2 pounds per acre, depending on the soil and the chemical. When distributed uniformly to plow depth, they rarely exceed one part per million. The actual contaminative rate is therefore low. Nonetheless, Sheals (1955) noted that BHC at low levels caused a heavy reduction in microarthropod numbers. DDT caused reduction in mites but an increase in the number of springtails (Collembola), apparently associated with a decrease in predator numbers. Depression of numbers in affected groups lasted through 17 months. The same pattern is not followed by soil flora. Although numbers of molds and bacteria fluctuate widely in soils treated with normal levels of insecticides, there seem to be no permanent effects on them. This conclusion is, however, based on single application (Bollen *et al.*, 1954).

Surface applications, which only incidentally work into soil, are potentially more serious because they are far more frequent and widespread. Still, no sustained imbalances seem to occur at low application rates such as used occasionally in forest-insect control. Hartenstein (1960) could show no significant decreases in mites and collembolan populations in forests treated with DDT and malathion at normal rates of application. Only at higher rates did he demonstrate transitory effect. Herein would seem to lie the danger to soil organisms. The most probable means of acquiring high concentration in soil would be accretions of chemicals from repeated applications. Only stable chemicals could accrue in this manner. The fact that they can is abundantly supported. Ginsburg and Reed (1954) found DDT present at concentrations of 35 to 113 pounds per acre in soil under apple trees. Those authors noted soil residues in many other crop situations as well. The lowest value found was 4 pounds of DDT per acre in potato soils. The important point is that DDT was never applied singly in the amounts reported in this study. I have not seen one study in which increments of accumulation were related to changing organismic character of the soil; yet the accretions occur and assuredly are widening in occurrence.

The foregoing description is a confined view—it assumes for purposes of discussion that stable chemicals remain in soil to produce whatever effects they may. In fact, they do not always remain. Chemicals may be leached out in runoff water, as previously noted. Treated seeds may be picked up by birds and mammals. These seeds are, in essence, poisoned baits, with their effects—possibly even death—dependent on the type of chemical and the amounts ingested. Finally, soil organisms containing chemicals may be removed by burrowing or surface-feeding animals. This situation is complicated further when soil organisms, most notably the earthworm, have the ability to store and accumulate chemicals in the tissues. The transfer of a stable chemical in soil, its concentration in a tolerant organism, and its ultimate distribution to a susceptible vertebrate is a most interesting sequence, and one with profound importance. Because of this importance it is described in detail in a section devoted to this subject (Chapter 20).

There are, then, three general methods by which toxic chemicals intrude into soil: general distribution by tillage throughout the top few inches of soil; spot treatment, generally near growing plants or on seed itself; and surface contamination usually resulting from above-surface insect control. All affect soil organisms, and all are potentially hazardous to surface organisms.

Extensive soil contamination as a method of pest control is relatively new. Possibly this fact could be offered to explain why so little is known of its general effects. This explanation is only in small part satisfactory. Serious gaps in knowledge derive from the emphasis placed by pest control technologists on the two indirect measures identified earlier. These measures describe a limited concern—effect on crop yield in a single growing season. Much less attention is accorded to longer-term effects on yield. In both instances the greater part of work on chemical side-effects is directed toward toxicity to the crop plant or to the possible contamination of products with chemical residues. The avoidance of legal entanglements potentially restricting sale is the motivation of the latter studies. In short, from the outset, soil-pest control procedures have been developed by technologists with little interest in the total biology of sustained productivity.

These are broad criticisms, which are in good part subjective. They can be reduced, however, to specific criticisms. Let me list a few that I feel are important.

Most important of these is the effect on soil fertility itself. The studies to date, which conclude the effects are none or inconsequential, are based on single growing periods and single chemical applications. Realistic as these experimental bases are for initial consideration, they by no means cover the entire range of possible effects. The most important of these other effects come from accumulating levels of chemicals in soil.

Ecological studies, which have as their goal full knowledge of soil biology, are now seldom conducted by control technologists. Yet such studies are most apt to lead to control methods that exploit cultural and biological alternatives. Only these methods are free of residue problems and of the need for repeated chemical applications.

The faunal simplification that is occurring everywhere aboveground is now being extended belowground. Of course, cultivation itself is the prime reason for this reduction in the variety of organisms. But when high concentrations of chemicals occur, more than target species are affected. Pest species are clearly hardy; many enemies of pests are not. We seem increasingly to be creating partially sterile environments favoring hardier species. Average numbers of pest species are higher under such circumstances, and outbreaks become more frequent. A pattern of this kind has been repeatedly observed in both surface and soil organisms. It is odd that control technologists seem not to appreciate the relationship between pest numbers and complexity of fauna.

It is equally strange how very little thought is given to one elementary fact—that surface organisms are sustained by those below the surface. The most obvious clue to this fact lies with pest insects themselves; their period of life in soil is transitory, an interlude in a life cycle inevitably concluding on the surface. To call an underground species a "pest" does not alter the biological fact of constant exchanges between soil and surface. A depauperate surface fauna will normally be paralleled by a depauperate one belowground. Moreover, chemicals applied belowground commonly return to the surface in biological carriers. The chemical may be on the body of a dying organism, or, if the chemical is stable and capable of being accumulated in animal tissue, it may be brought to the surface by an apparently healthy animal.

The practice of sowing seeds with toxic coatings is rapidly expanding both in area and in the variety of chemicals employed. Treated seeds are, of course, intended to kill or repel some target organisms, but their toxicity may not be so limited. Other seed-eating species may remove the seeds from soil (hence perhaps qualifying as a pest for so doing), or, if broadcast, a good proportion of seeds may remain on the soil surface, where they cannot germinate and are readily accessible to seed-eating forms. The latter likelihood is increased by aerial sowing, now practiced so widely. Whether in soil or on the surface, treated seeds are quite capable of harming species other than the intended target.

Lastly, it will unquestionably be only a matter of time before soil organisms which have been genetically selected for resistance to certain chemicals supersede nonresistant strains in the same way that many resistant surface organisms have. Continued exposure to a chemical is required to select for this kind of resistance. Exposure can be achieved by frequently repeated application, by high applications of a stable chemical, or by accumulating levels of stable chemicals. All three types of exposure are common and increasing. Resistance can be expected.

Terrestrial Invertebrates

Insects not only comprise the largest single terrestrial invertebrate class, but also as a group they are probably the most widely adapted animals on the surface of the earth. Only man is an effective competitor with insects. The bulk of pest control effort is directed against plant-feeding and disease-carrying insects.

There is no need to consider eradication of insects in general. Perhaps all that need be said is that "control," i.e., reduction in numbers, of the relatively few insects that we call major pests can be achieved only temporarily. Chemicals are the primary control instruments now in use. To affirm that they are effective for their limited purpose, one need only note the thousands of articles that have been published on insect control. The *Journal of Economic Entomology* contains dozens of articles per year attesting to successful reduction of particular insect species. The countless advisory circulars and bulletins of governmental agencies throughout the world further establish that insects can be controlled. The seven volumes to date of the *Annual Review of Entomology* have many articles summarizing progress in particular phases of insect control. The summary volumes *Advances in Pest Control Research* are devoted almost entirely to insects. Similar information on the methods and results of control appears in hundreds of other publications, and the number of articles is increasing at such a rate as to overwhelm a conscientious student of pest control.

My point is a simple one—in a limited way insect control is successful. I use "limited" to emphasize that the reduction in numbers is usually local and temporary. Most insect control leads neither to total removal of the pest species nor to reduced necessity for its control. The advantages of limited control are nonetheless great, and there is even indication that the numbers of insect control workers (and presumably their success in control) will continue to increase. Killing insects remains the chief preoccupation of applied biologists.

My focus in this book has been on the adverse effects of pest control. There are many kinds of effects, but, limited only to insects, problems of this kind reduce to three categories: the harm done to insects that yield useful products; the reduction in numbers of parasitic and predaceous insects effective in the natural control of insect populations; the failure to preserve insect species as objects of study or recreation.

The last of these is not taken seriously by many entomologists. Natural preserves for insects have been suggested by a few, particularly Europeans. Occasionally the idea is extended as an insect control measure. One serious suggestion was the establishment of small "islands" of natural fauna distributed throughout intensively cultivated lands. These islands would constitute reservoirs from which colonizing insect enemies could emigrate to surrounding depauperate farmlands. The idea is intriguing, but it must be noted that pest species would also survive in such a preserve, albeit in low numbers. Moreover, the preservationist's outlook, commonly applied to birds and mammals, has few proponents among applied biologists.

On the second point, beneficial insects that aid in natural control of injurious forms are often harmed by insecticides, with the result that insect control becomes more difficult. Frequent reference has been made to this problem in the text, and a detailed discussion is presented in later sections.

The first problem is acute. Insects of immediate commercial value to man are few. Lac insects, silkworms, and honeybees are examples. In this country the honeybee is singularly important, not only for its honey and beeswax but also as the major pollinator on croplands. Orchardists some fifty years ago began the practice of bringing hives of bees into groves of fruit trees. The practice expanded as increased yields in other crop types were demonstrated. The unusual value of bees in increasing seed yields in legumes was clearly shown during World War II. Since that time it has become common practice for growers to install hives of bees in fields during flowering periods. A profitable business has resulted from the rental of bees; bee-keepers no longer consider honey their chief source of revenue.

Unfortunately, bees are vulnerable to many organic insecticides. At the time when arsenicals were commonly used on crops, the chief hazard was to hive bees. Arsenic residues brought back to the hive were incorporated into honey, often poisoning entire broods of bees. Arsenicals are no longer widely used. The main danger now is contact toxicity to field bees from dusts and sprays applied to foliage. A lesser, though important, problem is contamination of places where bees drink. From either source, the hazard is real; heavy mortality has been observed repeatedly. The bee journals (for example, *Gleanings in Bee Culture*) have published many articles on insecticide poisoning of bees and precautions to observe in placing hives or timing insecticide applications. A few articles and editorials are strikingly similar in tone and content to the objections to pesticides raised by wildlife enthusiasts.

Most insecticides in wide use have moderate to high toxicity to bees (Anderson and Atkins, 1958; Todd and McGregor, 1960). Variations in timing, through observations of peak bee activity, and adjustments in dose rates can reduce the hazard inherent in some chemicals (for example, DDT, toxaphene, Trithion). Many other chemicals are highly toxic, and should not be used for several days preceding or during the pollinating period. In this group are such common insecticides as the arsenates, dieldrin, heptachlor, parathion, and Sevin. Instances of bee mortality continue to recur. Although both legal actions and strong precautionary recommendations have resulted in reduced contact of bees with toxic chemicals, I can see no way of avoiding some loss. Occasionally it will be serious. I suspect that the consequences of bee loss are greater than currently believed. As a broad question, who knows what the total impact of insecticides is on pollinating insects of all kinds? The mortality of honeybees on intensively cultivated lands can provide only a partial measure of the total effect of reduction in pollinating insects, even on croplands. It is clearly inadequate as a measure for noncroplands.

7 · THE COLD-BLOODED VERTEBRATES

Because cold-blooded vertebrates in general seem to be only moderately vulnerable to ingested poisons, they are able to store quantities of residues which are not lethal to them but which may be lethal to their predators (see Chapter 20). However, gill-breathers show great sensitivity to poisons which contact the gill surfaces. In addition mortality may result from the reduction in food organisms that is effected by pesticides.

Fish

Instances of local fish-kills are now so commonplace that they attract little local attention. In some kinds of control operations, fish loss is unpreventable and is an expected consequence of insect control. The usual sites of damage are the margins of lakes and streams adjacent to treated areas, major water-diversion channels through agricultural lands, and ponds or sumps where vector control operations take place. Fish may be replaced by stocking (particularly after vector

103

control), or reduction in numbers may be so minor as to constitute no permanent hazard to a resident population. Normally the reaction is topical and immediate. However, it is not always so. A more serious problem occurs when runoff waters are contaminated. Surface irrigation waters diverted into collection channels after use are commonly polluted. Certain crops, notably rice, require constantly flowing water. Insecticidal treatment on rice fields inevitably results in diversion of some toxicant from the treated area. Heavy rainfall following insecticide treatment may wash insecticides into ponds and streams. A slower process, but of the same character, is the leaching of insecticidal residues from soil. The insecticidal "contributions" from a major watershed or crop-growing area may result in considerable contamination a great distance downstream.

Fish mortality has been extensively documented in all the foregoing situations. (See, for example, U.S. Public Health Service, 1961.) The simple fact that such losses occur is not very important; they also occur from sewage and industrial pollution and from purely natural causes. Moreover, human demands on water supply have required impoundments, channels, and surface uses, which, taken together, affect fish populations more seriously than does insecticidal pollution.

Nonetheless, direct insecticidal effects on fish are important and are becoming increasingly so for these reasons. Fish are sensitive indicators of contamination. How sensitive is illustrated in Table 15. Fish mortality is an early and obvious warning that water quality has deteriorated. The necessary re-use of water dictates pure water.

Local losses of fish can be tolerated when, on balance, a greater gain than loss results from pesticide application and when natural or artificial replacement is possible. But "local" cannot be construed to include large bodies of water, entire watersheds, and large segments of streams. The recovery of fish populations under these conditions is neither rapid nor sure. Fish are an important resource in themselves. To tolerate occasional local loss is not to condone extensive loss. Local advantage cannot be traded for general disadvantage, even though the effects may be indirect and remote.

Perhaps more serious than the foregoing are the attitudes of some interests and their reflection in law. A prominent legislator, in a public hearing, told me that he represented agricultural interests only, and that as far as he was concerned water had only to serve crop-growing interests. He generously exempted urban and industrial users from his general indictment, but he was quite prepared to sacrifice all other water uses. I have repeatedly met industry and grower representatives with similar views. Neither the nutritional nor recreational users weigh heavily with these individuals. Aesthetic or moral responsibilities weigh not at all. Notably lacking is real knowledge of the complexities of quality maintenance in water and of the extent of re-use. In practice, many individuals intentionally disregard existing protective codes and do so knowing full well that prosecution is unlikely.

TABLE 15

Approximate median tolerance limits (TL_m) values to fish of several common chlorinated hydrocarbon insecticides and the practical application rates necessary to reach these values (from Henderson et al., 1960)

Pesticide	96 hour TL_m (parts per billion)	Pounds per acre applied to surface of water three feet deep to reach the TL_m concentration
Endrin	0.60	0.005
Toxaphene	3.5	0.03
Dieldrin	7.9	0.07
Aldrin	13.0	0.11
DDT	16.0	0.13
Heptachlor	19.0	0.16
Chlordane	22.0	0.18
Methoxychlor	62.0	0.51
Lindane	77.0	0.63
BHC	790.0	6.4

Less forgivable is the illusion, created in controlling legislation, that chemicals remain where applied. Residue laws allow a certain amount of an applied chemical to be removed from the area where applied; the amounts are spoken of as "tolerance levels" on foodstuffs. The implication is that no more than these amounts escapes from the area where applied. This is simply untrue. Contaminated runoff is one such avenue of escape. I will describe others in later sections. In all cases the premise is not valid that we are protected by residue laws from exposure to crop-protection chemicals. The best single negation of the premise is that frequent fish-kills result from pesticides, even when pesticides have not been applied in the immediate area.

As in most studies of economically important animals, effects of poisoning rather than causes are studied in fish. It is enough here to point out that the cause of rapid death in fishes exposed to chemicals is usually suffocation—interference with respiration at the gill membranes. Delayed effects, secondary poisoning, and disturbance of food-chain relationships are greatly different problems which have not been well studied. These subjects are considered later, in appropriate sections.

I feel it necessary here to substantiate that population numbers of fishes can be severely reduced by normal insect control practices, and that these reductions can affect a large area. The first example concerns aerial spraying for budworm in Canadian forests at 0.5 and 1 pound of DDT in oil per acre. The second illustration applies to warm-water ponds and streams in the cotton-growing regions of northern Alabama.

In each spring since 1952, large forest areas of northern New Brunswick have been sprayed with DDT in efforts to control a spruce budworm epidemic. The epidemic widened in spite of control efforts, and in 1957 some five million acres were sprayed, much of it for the second or third time. Fortunately, well before the epidemic occurred, personnel of the Fisheries Research Board of Canada had been studying Atlantic salmon populations in the major drainages of New Brunswick. A completely reliable "before" assessment could therefore be made (in contrast to most studies). Effects of spraying on salmon were immediately obvious, and led to continued and amplified efforts to assess the "after." Figure 1 graphically illustrates numerical reductions and effects on age classes in salmon of one river system relative to spraying operations near the river. Note that the effects are differential; ages and species are not affected uniformly. Additional effects also occur. Growth of parr is slower because of reduction of their food. Recovery of trout populations requires two years. The salmon are anadromous fishes normally spending three years in fresh water before moving to the sea. Under repeated spraying, the year-groups 0, I, and II can be affected more than once, and smolt production can be seriously curtailed. The number that return to fresh water has yet to be fully assessed, but there is a strong likelihood that salmon runs, as well as smolt production, will be severely limited (Keenleyside, 1959).

Only recently has it been possible to assess the long-term impact on salmon populations of continued forest spraying in New Brunswick (C. J. Kerswill, personal communication; Fish. Research Board, 1963). Earlier observations on immediate losses to salmon and aquatic insects have been repeatedly confirmed. Following aerial application of DDT at 0.5 pound per acre, predictable losses are 90 per cent in underyearling salmon, some 70 per cent in parr, and virtual annihilation of aquatic insects. Significant recovery of populations of aquatic insects requires five or six years. Commercial catches of salmon in marine waters have declined as predicted. The watersheds of the Miramichi and Restigouche rivers contribute approximately four-fifths of sea-run salmon numbers. It has been estimated that if DDT spraying in these

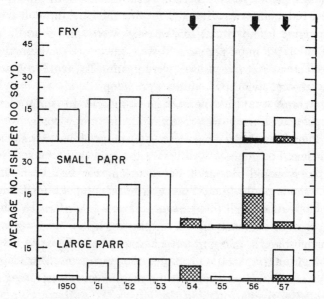

Fig. 1.—Abundance of young salmon in the Northwest Miramichi River. Arrows indicate years when DDT was sprayed in the area. Cross-hatching indicates fish found in sprayed parts of river (from Keenleyside, 1959).

watersheds had not reduced salmon numbers, commercial catches would have been one and one-half to two times greater than they have been since 1954. The commercial loss lies between one-half and one million pounds per year. Salmon returning to fresh water provide yet another measure of salmon loss. In the Tobique River, for example, direct counting revealed a drop in numbers in the decade 1953–62 to about one-sixth of original numbers. Angling success among sports fishermen has declined in proportion although fishing pressure has markedly increased.

Sportsmen, fishermen, and residents will attest to the recreational and economic importance of the Atlantic Salmon Fishery in New Brunswick. This is not a minor problem.

It has not been considered a minor problem by Canadian research workers. Close liaison has been maintained among affected disciplines; funds have been increased; helpful comments have been invited. These workers have sought sensible and realistic conclusions. With this foresight it is frustrating to report that a particularly well-planned program nonetheless yielded serious mortality of fishes and other aquatic organisms. The campaign was directed toward the black-headed budworm on Vancouver Island; about 155,000 acres were sprayed at the rate of 1 pound of DDT in oil per acre. Testing sites were established well in advance, and careful censuses were made. To avoid contact with streams, the following procedures were adopted:

(1) Streams would not be used as boundaries for spray plots. This would prevent streams from receiving double spray doses.

(2) Pilots would spray parallel to the course of major streams, keeping one swath away.

(3) Spray would be shut off when streams were crossed.

The hazards of the spray operation to fish and fish-food organisms were fully realized, and pilots made conscientious efforts to follow the above procedures. Nonetheless, mortality was heavy in most streams. In four major streams, the mortality of coho salmon fry approached 100 per cent; in others, mortality ranged downward to 10 per cent. The effect on salmon is best expressed by the original report (Crouter and Vernon, 1959: p. 37):

The coho salmon has a predominant three-year life cycle and very little overlapping from other year classes can be expected. Even assuming that there was not a complete loss of fry in the four grossly affected streams, and assuming also that there will be some smolt production, judicious management will not restore the population for many cycles. The economic loss will then be accumulative until the escapement is restored once again to the 1956 level.

The authors add that artificial stocking would be not only expensive but quite probably unsuccessful. The effects on fish-food organisms paralleled the effects on fish. The present unproductive state of the affected streams would not support a significant transplant. The effect on other salmonid fishes (resident trout and freshwater stages of steelhead) echoed the loss of coho fry.

Both the application of DDT and the study of its effects were conducted with great care. The contamination of streams during forest-insect control programs seems both predictable and unpreventable. One might well wonder what lies ahead for fish populations if aerial treatment of forests with pesticides continues and expands.

Pesticides need not be applied directly to waters to contaminate. As noted before, rapid runoff of rain and irrigation water can carry important amounts of stable insecticides (e.g., Bridges, 1961; Nicholson *et al.*, 1962). In 1950 almost all cotton fields in northern Alabama were treated in efforts to subdue a boll weevil outbreak. Some nineteen million pounds of insecticides were applied. Almost all insecticides applied were chlorinated hydrocarbons, characterized by high stability and toxicity to fish. The most commonly used insecticide was toxaphene, a chemical now frequently used as a fish poison. Unusual rainfall at the time of the outbreak necessitated retreatment of many areas. The rapid runoff of waters, with the unusually heavy insecticide applications, combined to produce a staggering loss of fish and fish-food organisms in thirty-six streams in the treated area. The contamination was felt over many miles of individual watercourses. Balanced productivity of the affected streams had not recovered even three years after the extensive mortality (Hooper and Hester, 1955; Young and Nicholson, 1951). Although these circumstances were unusual, the manner of contamination gives warning that it might easily recur. The increasing incidence of large-scale fish mortality should give clear evidence to all that chemical "control" is not sufficiently controlled.

Amphibians and Reptiles

Observations of pesticide effects on amphibians and reptiles are usually made casually in the course of other studies. Hence there is not a large body of data on which to base conclusions. Several good accounts of mortality are available, but these accounts rarely describe effects on the whole population. The permanent significance of this mortality can only be estimated.

In general, amphibians seem to be somewhat more sensitive to insecticides than reptiles. This differential susceptibility is presumably related to the character of the skin. Reptilian skin is heavily keratinized, an adaptation to terrestrial habitats. Respiration is entirely through the the lungs. The thin, permeable, heavily vascularized skin of most amphibians enables them to carry on dermal respiration in addition to respiration by other means. All amphibians are "aquatic" in the sense that they are tied closely to water for reproduction, and some, particularly salamanders, live constantly on a moist substrate. There is a wide range of dependence on water in both groups, however. Some reptiles are bound to aquatic environments more closely than are some amphibians.

The reptiles and amphibians which are adapted to live at the shallow margins of streams and ponds or which live on the exposed surface of the ground seem affected most by current means of applying chemicals.

The critical factors for the occupants of water margins seem to be three: lack of dilution of toxicant in shallow waters; the toxic character of diluents and solvents used with pesticides, especially as they pass through the permeable amphibian skin into the vascular system; and the tendency of oil-based pesticides to concentrate at the water surface. The last point—concentrated environmental exposure—is, I believe, particularly important. The water surface is repeatedly broken by frogs and certain salamanders. The aquatic snakes move primarily at the surface. Even the turtles, which are seemingly quite resistant to pesticides, frequently break the surface to breathe or sun themselves.

Ground-dwelling reptiles and amphibians contact surface contaminants in a similar way, but the contact is less apt to be intermittent. On the other hand, arboreal forms or those that keep well hidden do not maintain continued exposure to surface-applied chemicals. It is well established that chemical applications are "filtered" by heavy overlying plant cover; very small amounts, if any, reach under logs and stones or into the subterranean crevices where many species dwell.

The foregoing conclusions are inferential. They are, however, based on known instances of mortality. Most examples derive from experience with DDT (see Rudd and Genelly, 1956). With this chemical it appears that mortality in amphibians is severe at an application rate of 1 pound per acre and becomes catastrophic at 2 pounds or more per acre. Fashingbauer (1957), for example, records almost total destruction of a population of wood frogs (*Rana sylvatica*) in an area sprayed at 1 pound per acre for tent caterpillars. At the lower rates, toxicity seems largely dependent on solvents or the particular kinds of environmental exposure I have described. Comparable responses in reptiles begin at about 2 pounds per acre, probably becoming serious at about twice this rate. Population effects are noted less commonly. Logier

(1949) predicted a 50 per cent mortality of reptiles and amphibians where 1 pound per acre (DDT in oil in forest-insect control) reaches water surfaces; he concluded that such a mortality would not be serious if not often repeated or if not widespread. Stickel (1951) concluded that populations of box turtles in a woodland area in Maryland were not visibly affected by annual applications of DDT at 2 pounds per acre.

The fire ant control campaign in the South provides the only other good source of data on pesticidal effects on reptiles and amphibians (DeWitt and George, 1960). In this instance the chemical concerned was heptachlor in granular form at 1.5 to 2 pounds per acre. Unintended contamination of water resulted from faulty application and (probably) from runoff of rain waters from treated land. On one study area in Alabama, three genera of snakes disappeared within about a year's time; one purely terrestrial form survived in low numbers, but its reproductive patterns were affected. A similar but less severe pattern was observed in snake mortality in two parishes in Louisiana. Both aquatic and terrestrial turtle populations were severely reduced, and show no signs of recovering. Lizards seem similarly affected. Four species of the frog genus *Rana* virtually disappeared from one study area. Arboreal species of frogs and salamanders were little affected. There is no question that the effects of this control program on reptiles and amphibians were catastrophic.

Delayed or indirect effects may also occur in these groups, but discussion of them is reserved for later sections.

8 · THE WARM-BLOODED VERTEBRATES

Inadvertent effects of pesticides on birds and mammals have received more attention than have the effects on lower animal forms, but even so, the documentation is meager and incomplete. Much of the mortality is the result of secondary poisoning which, because of the absence of circumstantial evidence, is difficult to attribute to pesticides without analysis of body tissues.

Mammals

Information on the toxicity of pesticidal chemicals to mammals comes from three sources: intentional control of pest species, chiefly rodents and carnivores; intentional use in toxicology and medicine to assess hazard to human beings and livestock; and field studies on wild mammals exposed unintentionally to pesticides (Rudd, 1961). The intentional uses of chemicals against mammals are the subject of a voluminous literature and will not be stressed here. It should be noted, however, that the greater part of data on mammalian response to chemi-

112

cals comes from these sources. In fact, the use of laboratory animals in the bioassay of pesticides is the primary reference base for assessing the toxicity of these chemicals to human beings and domestic animals (see Table 2). Furthermore, most regulatory judgments are based on information of this type. For pertinent summary information, see Negherbon (1959) and Rudd and Genelly (1956).

Emphasis here is on the unintentional effects of pesticides on wild mammals and on the effects of mammal poisons on species other than mammals. Unfortunately these emphases concern the least documented areas of pesticidal effects on mammals. This is true, and apt to remain so, because the wild mammals of North America generally do not have the intrinsic popular appeal possessed by some other animal groups. More frequently they are looked upon as nuisances, disease vectors, objects of fear or superstition, or economic liabilities. Research investigations on animal–pesticidal interactions rarely center on wild mammals, whereas they do commonly for insects, aquatic invertebrates, fishes, and birds. Noteworthy exceptions are the game mammals and, in a few connections, the fur bearing forms.

Unintended mammal deaths result largely from (1) poisoned baits used to control nuisance mammals, (2) abnormally heavy applications of chlorinated hydrocarbon insecticides, or (3) special environmental situations (usually at lighter rates of pesticidal application) where greater exposure is rendered likely.

Toxic chemicals intentionally placed on acceptable foods (poisoned baits) will always be hazardous. They are so because mammal poisons are as a group highly toxic, and, equally important, baits are attractive to many kinds of animals other than the target species. In many instances careful placement of baits hinders their accessibility to non-target species (in burrows, for example). Countering this precision, however, is the increasing use of aircraft in dispersing poison baits.

Control of wild dogs (coyotes, wolves, and dingoes) with poisoned flesh baits has always resulted in some loss of other species of carnivorous mammals (skunks, foxes, weasels, domestic dogs). There seems to be no way of preventing losses of this kind. The campaigns to control rabbits in Australia, deer and opossums in New Zealand, and field rodents in this country have produced unintended mortality. The actual extent of loss is not known. The only group likely to know its extent, or in a position to learn it, is that responsible for the control operation, but reports are incomplete or nonexistent. However, furbearing and game mammals as well as domestic dogs have their vociferous defenders. As a result, wherever these mammals are poisoned or endangered, controversy appears. No mammal control campaign of this character has escaped it (New Zealand Forest Serv., 1958; Robinson, 1953; Rudd and Genelly, 1956; Smelser, 1959).

Poisoned mammal baits may have birds as their unintended victims. Direct poisoning depends, of course, on food predilections (secondary poisoning is discussed in a later section). Flesh baits have killed magpies, crows, and eagles, but apparently this loss is not extensive. Grain baits probably produce the greatest unintended loss of this kind. Graminivorous birds have been found dead on many occasions in areas undergoing control. In 1960, for example, 1080-treated grain bait distributed by air for forest rodents killed several species of birds, with heaviest mortality of Steller jays (E. G. Hunt, personal communication). In this instance no real searches were made for bodies; they were found at camp grounds. Thousands of geese were killed in the Tule Lake area of northern California and Oregon by carelessly distributed grains directed at meadow mice (Mohr, 1959). Repeated losses occur among pheasants and quail under similar circumstances.

Abnormally heavy broadcasts of nonselective insecticides, such as the chlorinated hydrocarbons dieldrin, DDT, endrin, and heptachlor, are capable of killing mammals. In insect control programs these chemicals will normally be applied at rates of 0.25 to 1.5 pounds of toxicant per acre. Within this range no lasting direct effects on mammals have been noted. Where, however, the special habits of a pest insect require high insecticide doses and long residual action, mammal mortality has been common. In a campaign in Illinois against Japanese beetle, dieldrin was used at 3 pounds per acre. Ground squirrels, muskrats, rabbits, and other vertebrates were killed (Scott *et al.*, 1959), and general declines in population numbers followed because of continuing direct loss. The program against the imported fire ant in the southern states was based initially on 2 pounds of dieldrin or heptachlor per acre (dose since lowered). Exact figures on loss are not available, but populations of some species declined markedly (e.g., raccoons, rabbits). Mammal mortality was revealed in all investigations, and many specimens examined contained chemical residues in tissues. Surviving raccoons in one area studied still had residues in their tissues a year after treatment (DeWitt and George, 1960). DDT, below 4 to 5 pounds per acre, does not affect mammals directly, but indirect effects have been noted at even normal insect control levels.

High levels of application cannot be called "abnormal" when they are designed to kill mammals. Application rates of these chlorinated hydrocarbon "insecticides" to control mammals range from 1.5 to 6 pounds per acre, depending on chemical and formulation. All such chemicals can and have killed a great variety of vertebrate animals. Needless to say, insect populations are also drastically reduced. Endrin, the most toxic of those used, is not recommended by conservation agencies, but is commonly recommended by others, particularly for mouse control in orchards. The hazard to wildlife, pets, domestic animals, and spray operators is high enough to warrant discontinuation of this practice.

Lastly, we may look to those hazards that depend on a particular kind of exposure. They may result from the behavior peculiar to a species; for example, deer, to their misfortune, are attracted by the "salty" taste of sodium arsenite, used in forestry to remove tree bark. In other cases a chemical, although lightly applied, concentrates greatly at a particular site. DDT applied to the foliage of elm trees concentrates at the crotches of major branches and in the soil at the drip line of the tree. Other particular environmental features may trap or concentrate the chemical; such seems to be the case with applications of DDT in oil on muskrat marshes. The oil solvent containing insecticide floats at the water's surface. Although immediate toxicity may be discounted, some longer-lasting effect, reflected in population declines, seems to be operating (Stearns *et al.*, 1947; Wragg, 1954).

To summarize, direct toxicity to mammals from normal insect control operations is probably rare. Operations requiring relatively high application rates are hazardous to mammals as well as other forms of life. Mammal control operations are intrinsically hazardous. This inherent hazard can be ameliorated only by confining the exposure to the target species by exploiting its distinctive habits. Indirect effects on mammals are probably greater than presently known. Some known effects are described in later sections. Irrespective of cause, pesticide effects on mammals will continue to incite contention.

Birds

No other group of animals is as popularly attractive as are birds. They, like the fishes and the moths and butterflies, have evolved a complex of adaptive types that are colorful, ubiquitous, and, most important, noticed. Birds are the first identification with natural free-living organisms that we have as children, and throughout our lives their presence is closely identified with human activities and judgments. Birds are a valued part of our culture and will remain so. The organized "bird watchers" are a very small proportion of our population, but they are far from representing the total interest in birds. None but the most sadistic of persons could want to kill birds unnecessarily. In those instances where planned killing is sanctioned, legal codes exercise every restraint consistent with need.

For these reasons, losses of birds from chemicals have drawn a great deal of attention. It is a disproportionate attention from a purely biological view. Many other groups of animals may be as affected by chemicals, but they have no defenders. Human values are biologically disproportionate, and, this being so, no apology need be made for our special concern for the welfare of bird life.

The habits of birds dictate that they shall meet chemicals in many places. There is great variety. Cropland, forest, hedgerow, sea, stream bank, pond, river—each environment supports its distinctive birds. There are ground-dwelling and tree-dwelling species; burrowers and soarers; urban and rural types. Therefore, wherever chemicals are used outdoors, they can contact birds.

The fact of contact does not necessarily mean harm. Birds are not vulnerable to most chemicals in normal use, nor are all species in a treated area apt to be affected at any one time. The chief protection of most birds is their mobility. When able, they move from an area undergoing chemical treatment. Anyone who watches a mist blower or dusting plane in operation will observe the flight of birds from the treated area. Since the spraying operation is normally maintained for some time, birds do not re-enter an area quickly. This is particularly true of passerine birds whose food habits are not narrowly restrictive.

Some kinds of birds are restricted to narrow habitats and are driven from their normal habitat only with great difficulty. The wading birds are good examples. A pond, marsh, or rice field is their home. It matters very little that a stubble field, orchard, or alfalfa field is nearby. The fact of habitat specialization ensures that treatment of such areas will cause mortality if the chemical and its vehicle inherently possess the potential.

The loss of egrets in California rice fields is a good example (Rudd and Genelly, 1956). Dieldrin at recommended doses is toxic enough to kill wading birds. Egrets cannot be driven from a field. They simply move to another part or, if available, to an adjoining slough or drainage ditch. This is scant protection, since aerial applications are not so accurate as to avoid contamination of the field boundaries. The precise manner of injury to species of narrow habitat proclivities is largely an academic issue. The egrets, for example, are probably injured through a combination of direct and secondary intoxication. Constant dermal contact with the toxic surface film unquestionably has its effects. Less sure, but not to be ruled out, is the contamination of fish and other aquatic organisms that constitute an egret's food supply. But, whatever the means of causing injury, the immediate consideration is the restricted environmental requirements of a species. Technical information can be useful in determining methods of reducing exposure, but it does not alter the basic requirements of the species or their susceptibility to exposure. This is a point curiously ignored by most professional toxicologists.

Valuable as is the mobility of birds, it cannot be entirely depended on for protection. In addition to narrow habitat requirements, the cyclic or rhythmic habits of birds also counter mobility. The result of cyclic behavior patterns is that the bird is forced to remain in a given habitat for a period of time.

The most obvious of these in the daily cycle is roosting. While mobility protects birds from immediate exposure to chemicals during daylight hours, their return at night to a favored habitat is a rule, not an exception. If the foliage is wet and if chemical residues are not only toxic but available, or if applications are made at night, birds, denied their diurnal mobility, are exposed in full force to the toxic capability of the chemical in use. Loss from this type of exposure, although difficult to document, is, I am quite sure, widespread. It is known to occur in passerine birds in citrus orchards sprayed with parathion, where the protective canopy of vegetation apparently serves also as a fumigation umbrella. It is known to occur in such closely lying plant structures as alfalfa and ground covers under orchard trees. Pheasants and other gallinaceous game birds have drawn the most attention in these instances. But many passerine birds have also been killed in roosting sites.

In intentional control this type of hazard can be effectively employed (Janzen, 1961). Reducing the numbers of crows, starlings, and blackbirds depends in part on treating roosting sites; because these birds frequently form large flocks, the combination of large aggregations and restricted habitat at roosting renders them particularly susceptible to control. The success of these treatments, however, should be taken as a measure of hazard to desirable species.

A second restriction on bird mobility is that imposed by the usual specificity of their food. Few birds approach the omnivorous eating habits of bears, skunks, or pigs—or man, for that matter; the manner of acquiring food is restricted for a given group or species of birds. The foliage-feeding insectivorous warblers and vireos, for example, cannot substitute fruits or seeds or grasses for insects. They range intimately into the parts of plants where feed the insects, mites, and spiders that are their food and that also are the target of a chemical application. Birds and economic entomologists are directed toward the same ecological niche, but the entomologist has the advantage in this competition. Not only can he reduce a food source for birds but his manner of doing so frequently involves chemicals capable of harming the birds directly. There is no escape for the birds if the niche is treated.

This food predilection is not limited to foliage-feeding birds. Ground-feeding birds will be largely insectivorous or graminivorous. Although different kinds of food within a food class may be eaten, these differences are only seasonal adaptations and do not amount to great differences in habit. Temporary increases occur in the numbers of carnivorous birds in areas of rodent abundance, but not so regularly as might be supposed. And the type of food is the same even though the particular species may not be the usual prey. The only significant differential is the condition of prey. For my purposes here, "condition" refers to the numbers of individuals in a population and to the well-being of the individuals in a population. And both conditions modify the ease of capture by carnivorous birds. If numbers are high, the likelihood of capturing prey is increased. Carnivorous birds are well fed in years of rodent abundance. Breeding is successful, and even slightly more successful than the norm if food is readily available throughout the reproductive season. When numbers of rodent prey are reduced, capture is less assured and carnivorous birds face a real task in merely remaining alive. Starvation is a common event among predaceous birds and mammals, and in some instances is dramatically noticeable, such as the winter die-offs of owls recorded in Germany in recent years. Needless to say, the carnivorous birds that survive in years of low rodent numbers do not breed successfully, for degree of success is related to the ease of capture of prey. In some years, breeding will begin but not be completed, and little production of young birds occurs.

The second "condition" related to ease of capture—prey well-being —concerns poisoned prey. Secondary poisoning will be discussed in a later section devoted to the subject; here, food alone is being considered. A classic argument derived from Darwinian controversies is the nature of prey taken by predaceous animals: Do predatory animals take "unfit" animals—the diseased, sickly, and unnaturally exposed? Darwinism has been abused in many areas. This is one example. It has been shown in many studies that predators in fact prey primarily on the prey animals that are easiest to capture. But whether our predator is falcon, wolf, or black bass, it should be remembered that these animals can take perfectly healthy animals. Such animals are simply more difficult to capture in periods of low numbers.

I have wondered why secondary poisoning among raptors occurs more frequently than one might expect on a random basis (Rudd and Genelly, 1956). If a rodenticide—necessarily a stable chemical (e.g., thallium)—poisons many rodents in a population, the likelihood that poisoned individuals will be captured certainly increases. An animal in convulsions is hardly a "fit" animal in the Darwinian sense, and without doubt such individuals are selected differentially. It is difficult to understand otherwise how such large numbers of owls in Germany contained thallium in tissue following recent rodent control campaigns. A widespread rodent control effort is rarely as effective in reducing numbers as advertised, nor is the control effort normally applied uniformly to include the feeding territories of many individual carnivorous birds. I can only conclude that in areas where rodent poisoning is conducted, carnivorous birds will find that poisoned animals are easy targets, and consequently secondary poisoning will be likely.

Just as the environmental mobility of birds is restricted by food and roosting, so also is it restricted by the habits of reproduction. Annual nesting and the rearing of young increase the vulnerability of birds for a quarter to a third of the year. The rule with most birds is confinement to a territory, i.e., a spatial area normally adequate to sustain both adults and young, through the reproductive period. Normally, adult birds are tied not only to the nest and young but also to the territory. However, even slight disturbances may often result in abandonment of nest and young. The contamination of a territory with a toxic chemical can poison adults and young birds or lead to nest abandonment because of disturbance. The former result depends a good deal on the nature of the chemical, and also somewhat on the diluent or vehicle in which the pesticide is dispensed. Even water sprayed on nestlings causes such a temperature loss by cooling that death may result. Thus, any spraying may injure nestlings if it contacts them directly. Nests are normally hidden and well protected, so that such loss is not likely. Mortality of this kind is most apt to result from orchard spraying. Forest spraying as normally practiced does not result in disturbance sufficient to cause abandonment of territories, nests, or young. Nor does it through toxic action seem to harm either adults or young directly. Specialized operations (as examples, Dutch elm disease control, Japanese beetle control, and tick control) and experimental operations, both of which may demand high dose rates, have, however, had marked effects on nestlings as well as adults.

The usual index to bird numbers is a census based on sight and sound. During the nesting season, numbers of birds are easily estimated by such a census. Not only are birds spatially limited but this also is the period of maximum singing. The "before and after" spraying censuses conducted by trained ornithologists can be depended on to give good relative counts, and in many situations may be taken as nearly absolute. During the nonbreeding period, however, bird mobility can be so great as to make counts of little precise use. I mention the limits of censusing methods because people untrained in biology frequently dismiss all bird-census results as meaningless. The reason usually given is the lack of precision. A truer reason is perhaps the convenience of the argument.

Different studies of pesticide effects on birds do not always reach similar conclusions even though types of chemicals and rates of application are the same. Variability of this sort will hardly raise a biologist's eyebrows, but it is a source of surprise to many physical scientists. There is a host of ecological variables, many beyond analysis (one is the acuity of the observer). It is not my intention to discuss these variables here. Rather I wish to state only one conclusion. Insecticide applications are certain to have ecological effects; after all, the purpose of application is disturbance (reduction of pest numbers).

The effects of DDT on birds are best known. DDT will cause adult mortality in birds in any ecological situation at 5 or more pounds per acre. At 2 or 3 pounds per acre it will kill adults under certain conditions, will kill considerable young, and will essentially eliminate the flying-insect population on which many insectivorous birds depend. At 1 pound per acre, direct effects upon birds are negligible (except where directly sprayed: viz., nestlings) but indirect effects will result from reduction of insect numbers. At rates under 0.5 pound per acre—the normal range for mosquito control—effects are indirect through reduction of food insects. These effects are noticeable, albeit transitory, but can be considered unimportant if chemical treatment is not repeated (George, 1960a; Rudd and Genelly, 1956).

These conclusions are formed from our experience with DDT. Other chemicals are relatively little known. There is no reason to suppose that the pattern is not the same if the necessary corrections are made for the inherent characteristics of the chemical. Dieldrin and heptachlor, for example, are more toxic than DDT. Where we have data, it appears that we can apply the same conclusions if we reduce the dose figures to one-half or two-thirds. Parathion, in contrast, is much more toxic. We would have to lower the values by three-quarters or more, but remember that the opportunity for effects to occur would be of shorter duration because the chemical is less stable.

In the first portion of this section I discussed the habits of birds that render them particularly susceptible to chemicals. The cloak can be reversed to expose its lining. How are chemicals applied in nature in a manner that endangers birds? There are both intentional and unintentional effects.

Birds as competitors for food or as nuisances must, on occasion, be controlled. Relatively few species of the many thousands in the world are continued targets of control. In almost every case the target species are truly commensal—live near human habitation—or have found agricultural environments more suitable than their native one. Familiar examples are crows, starlings, certain finches, and blackbirds. These adaptable species have doubtless increased their numbers—thereby qualifying as pests—because of man's activities. Their increase was produced unintentionally by man, and intentional efforts to reduce them are justified. Another group of birds may be considered here. These are the occasional competitors—for example, the crowned sparrows—which under some conditions are responsible for losses of seeds or growing plants. Here again, few species are involved, and only these species are the intended targets of control measures.

Mechanical devices for trapping and repelling birds are used locally to control birds, but only one system of control has general significance: the use of poison baits. Seeds and fleshy fruits are the normal vehicles for poison, although specialized baits (for example, suet for magpies) are used to suit the habits of a particular species. Standard chemicals such as strychnine are used normally, although chlorinated hydrocarbon insecticides have been tried in special situations. Blackbirds, crown sparrows, finches, and crows are baited regularly. In California and some other states, many thousands of birds are killed annually by such means. Chloral hydrate—the "Mickey Finn"—has been used in Sweden and France, and has the advantage that it merely drugs birds with low doses so that they may be removed to other areas.

One may question the justification for many of the instances of this kind of control. But, beyond this question, it is probably fair to say that bird control is reasonably well conducted to prevent unnecessary loss. Hazard to other species nonetheless exists, and actual loss of desirable species has been documented. Though baiting is the most widely used control system, another method, potentially more hazardous, is being developed. Uniform application of acutely toxic sprays whose persistence is short (for example, the highly toxic TEPP) has been successful on blackbirds. This method is applicable only to species that concentrate at nesting or roosting sites. The method is, of course, indiscriminate since other animals in the treated area will surely be killed by the doses needed to kill the birds. In this respect, treatments of this kind are similar to broadcast rodent control methods, already widely used and indiscriminate in their effects.

By far the greatest source of hazard to birds is unintentional intoxication. The extent of loss, I believe, is much greater than is indicated by detailed studies and published reports. This I infer for a number of reasons:

(1) Reports of damage to game birds in contrast to song birds draw a disproportionate amount of attention.

(2) Control biologists do not weigh heavily the loss of nongame species, if they consider them at all.

(3) Many of the disrupting effects on birds are either subtle or delayed.

(4) Reports from other countries verify the universality of chemical effects on birds.

(5) Commensal and other adaptable species have increased in numbers around dwellings, thereby giving the impression that there are "plenty of birds."

I indicated earlier how the habits of birds lead to their exposure to control agents. It remains now to consider how birds may actually contact chemicals. In general it resolves to this: If the chemical is applied to an edible substance, intoxication through feeding is the only possible entry into the body; if the chemical is not limited to a food substance (usually a spray preparation), contamination through the mouth, skin, or lungs is possible, and all may occur simultaneously. Only the habits of birds and the crop type dictate the routes of contamination, but, in general, multiple contamination is the rule.

Poisoned baits are used to kill cutworms, flies, wireworms and other larval forms in the soil, grasshoppers, carnivorous mammals, and rodents. The baits may be pieces of fruit or sweet concoctions, flesh, grains, or treated seed. The foodstuff itself is not intended to be eaten by birds, but it is. Examples of consumption of poisoned seeds are many. Karl Borg (1958) has reported extensive loss of game birds in Sweden from eating seeds treated with mercury compounds or aldrin. In this instance, seed treatment was intended to control fungus or wireworms. Similar losses in England have recently become a matter of national concern (Ministry Agr., Fish., and Food, 1961). I have described considerable mortality in pheasants and ducks in the rice-growing areas of California from ingestion of DDT-coated rice seed on the edges of fields (Rudd and Genelly, 1955). Treated seed is by no means inevitably exposed to birds. For example, seed corn in the Delta region of California is regularly treated with a fungicide or insecticide or both. In sufficient amount, either chemical could harm pheasants, an important game bird in the area. In practice, however, no harm is likely, because sprouting corn is husked by the birds before they eat it.

Grain baits treated with strychnine, thallium, zinc phosphide, or Compound 1080 are used to control field rodents. All such baits are also capable of killing birds. That significant loss actually occurs is attested by reports in recent years from the United States, France, Germany, the Netherlands, and Poland. A good deal of controversy has arisen over actual or presumed loss from this kind of baiting (summarized in Rudd and Genelly, 1956). Hazard may be minimized by selecting grains not normally taken by desirable species of birds in the area under treatment. Whole oats, for example, are rarely eaten by quail, although readily eaten by rodents (Emlen and Glading, 1945). In practice, strychnine treatment itself constitutes rather little hazard to gallinaceous birds; as a group, these birds are relatively resistant to strychnine. Doves, in contrast, like all columbiform birds, are particularly vulnerable to strychnine. Through the ingenuity of E. R. Kalmbach, the recommended procedure in preparing baits now includes coloring grain baits. Coloration normally causes birds to reject treated grain. The method is not infallible. The thousands of ducks and geese killed in the Tule Lake area of California (under recent intensive treatment for a mouse irruption) included many with colored grains in their crops; however, the majority were killed because an appreciable percentage of the mouse bait distributed on open fields was, contrary to recommendation, not colored (Smelser, 1959).

Flesh baits to control predatory mammals (coyotes, wolves in North America, dingoes in Australia) have caused loss of magpies, crows, and eagles. True carrion-feeders—the vultures—are able to regurgitate poisoned flesh quickly; it is probable that they are little affected by flesh baits.

The poisoning of birds by pesticides is common and increasing. It occurs because birds are found wherever poisons are used. It is increasing for three reasons: greater use of broadly acting chemicals, less discriminately applied; the expansion of seed-treatment practices; and the accumulation of toxic chemical residues in favored foods of birds. The last source of poisoning is so serious that I have given it special attention in a later section (Chapter 20). Whatever the control practice harming birds, concern for their welfare will remain because of their value as natural enemies of pests and because of the esteem which man traditionally has for them.

9 · THE SIGNIFICANCE OF REPULSION

By definition the word "repellent" must be confined to objects or chemicals that do not kill but are offensive in some way to an attacking animal. Repellents therefore do not include physical barriers or lethal poisons. They do include frightening devices; offensive tastes, sounds, colors, and odors; and in some instances sublethal "educating" toxicants. Their effectiveness is based on exclusion rather than reduction in numbers. Chemical repellents are normally applied to surfaces to prevent access, in which case their effectiveness is limited to a narrow band at or near the treated surface. Less commonly, a repellent may be effective throughout a relatively large area, not necessarily being focused on a particular surface.

Beyond these statements, generalization is difficult. The sensory receptors of animals are intriguingly diverse; so also are the manners of contact between attackers of man and his goods. And so too are the values justifying the use of repellents! An effective repellent is therefore one that has successfully threaded a specific pathway through these quite different areas of diversity. In other terms, our successful repellent has interfered with the normal sensory response of an animal; it has been put to use in a specific place to accomplish this; and it was selected because no effective alternative existed. On the last point, the repellent quite frequently protects more than one object, more than one value.

There are only three groups of animals for which repellency can be considered significantly effective in pest control.

The first of these groups is insects. Almost all successful effort to use insect repellents has been directed toward blood-sucking insects. They may be nuisance insects or disease vectors. The demands of military operations have occasioned a good deal of research leading to the development of low-toxicity, nonirritating chemicals that, applied to the skin, repel many different insects for hours. The number "612" is certainly familiar to soldiers, hikers, and others who move in the mosquito's environment. Repellents may also be useful in moth-proofing and in fly control on domestic animals. In all cases, however, the material acts at a surface and in no case can be considered a hazardous contaminant in nature.

Insect repellents are not widely exploited to protect crops, and there seems to be no practical likelihood that they will be (see Dethier, 1956, for review).

On the other hand, field use is applicable for mammal and bird repellents, the only other groups of repellents useful in pest control (Thompson, 1953). In almost every case where repellents are used to limit depredations from warm-blooded animals, both a product and the animal itself are being protected from potential harm. The only important exception is the prevention of damage to stored and packaged products by rodents, where the animal is not valued. Since damage of this kind does not apply to the field, I shall dismiss it here with but one comment: The majority of work on this phase has been conducted by the Wildlife Research Laboratory of the U.S. Fish and Wildlife Service; refer to Welch (1954) for an able review.

Most problems in agriculture and forestry in which mammalian repellents figure derive from the depredations of herbivorous or graminivorous species. More specifically, among herbivores, use of repellents has been to reduce damage to orchard and ornamental trees by deer and rabbits. (Rodents in these environments are better limited by lethal poisons.) Repellents for these species depend on offensive taste, and are sprayed or painted on plant surfaces. Several formulations are effective if heavily applied, but none as yet repels for more than a few weeks. Moreover, costs are relatively high. The "ideal" repellent for these species has yet to be found. It should be inexpensive, lasting, taste-dependent, and nontoxic (Thompson, 1953).

The requirements need not be quite the same in forestry. Sublethal toxicity is currently depended on to a great degree. Deer and rabbits cannot be repelled as effectively in this environment as in croplands, but (contrary to some opinions) depredations are not so great as to justify much expense. The chief problem—and a serious one—stems from rodent damage to seed in reforestation programs. Some years ago a program of seeding Douglas fir on logged land on the Pacific Coast had to be abandoned because forest rodents ate all the seed. The hazard to pine seeds is somewhat less, but still important. Direct seeding is usually an unsuccessful approach to the re-establishment of conifers. Planting seedlings is too expensive in the present economy. Some modification of seeding that would ensure adequate germination and protect seeds from depredations is therefore desirable. Reduction of rodent numbers over wide areas is both costly and temporary. Treatments of seeds or the ground area around them with true repellents have not fared well. Toxic seed coatings have proved effective, although hazard is present. However, the use of seed coatings of endrin, tetramine, or Compound 1080 at levels sufficient to sicken rodents but let them survive (perhaps killing some) has provided an important new approach to reforestation (Tevis, 1956). Once sickened, the rodent is conditioned to avoid both treated and untreated seed. Leaving the resident mouse population largely intact precludes an influx of nonresident, unconditioned mice. (Small mammals have some value in forestry as very effective predators of insect larvae in soil.) Learned avoidance is perhaps a better term than repulsion; it is not a true repellency, and there is some inherent hazard. Curiously, there is some indication that young mice are somehow taught to avoid treated seed. The effect of sickening thereby extends beyond the parent generation. Neither the reliability nor the extent of this phenomenon is established, but, if consistent, it will be significant in reforestation. Possibly future work will emphasize nonlethal chemicals capable of inducing nausea or other temporary ill effects, thereby reducing hazard.

There are two other kinds of manipulation that reduce access to desirable plants on rangelands or forests. One is to remove, by chemical means, favored food plants, discouraging high numbers of herbivores by destroying their food. This would be applicable to range pasture of mixed plant species. The other is diversionary: adding easily available supplementary foods at sites requiring protection. Neither of these methods can be widely applied as mammal control measures.

Both electric barriers and ultrasonic sound may be termed "repellent" in some situations, but neither is currently used in the field.

In review, acceptable mammal repellents depend on offensive taste or sublethal toxicity. Deer and rabbits can be repelled by the former means if cost can be justified. The depredations of forest rodents can be effectively restricted by the second approach.

Repelling birds is difficult. Frightening devices are the most widely used, but their effectiveness is limited. Numerous devices based on noise or glitter have been tried, but they are effective only for short periods at best. Frightening devices are sometimes ludicrous. The painted sheet-metal "owls" that the city of San Francisco purchased, for example, affected people and politics a good deal more than they did pigeons. The broadcasting of recorded distress calls, as developed by the Frings at Pennsylvania State University (e.g., Frings and Frings, 1957; Frings and Jumber, 1954), has some merit. Here too, however, birds usually become quickly acclimatized, though starlings can be discouraged from buildings by these calls.

Seeds are sometimes treated with toxic chemicals to repel birds (Meanley *et al.*, 1956). This is better avoided because most such chemicals depend on sickening birds, not truly repelling them. In general, birds have a poorer ability to taste and smell than mammals; hence, true repellency based on these sensations is not normally attainable. I was able, however, to reduce pheasant depredations on corn with sublethal levels of lindane applied either to seed or to the ground surface. Unfortunately the chemical concentration demanded exceeded the cost of corn damaged. This could hardly be considered a fair exchange. Also, toxic hazards mounted directly as chemical concentration increased.

The most promising approach rests on the visual acuity of birds. Many kinds of birds are extremely sensitive to colors, shapes, and textures. The only significant exploitation of this acuity is with colored rodent baits. Whereas rodents are essentially color-blind and pay little or no attention to the color of their foods, birds avoid unnaturally colored foods. The greatest aversion by birds is shown to colors near the center of the humanly visible spectrum—yellow and green. Dyes applied to seeds should make them effective repellents in most instances. The practical applicability of color repellents for birds was developed by E. Kalmbach, former director of the Wildlife Research Laboratory of the U.S. Fish and Wildlife Service (Kalmbach and Welch, 1946). The idea was imaginative and has partially resolved the difficulty of controlling field rodents while guarding birds. A movie with the engaging title of "Birds, Beasts, and the Rainbow" has since been widely circulated, and grain baits for rodents put out by state and federal agencies are now normally colored yellow, green, or black. The impression of safety, still largely current, is in part illusory. Dambach and Leedy (1948) found that pheasants were attracted by some green-colored "repellents." The recent loss of ducks and geese in the Klamath Basin, in part from broadcast colored rodent baits, was heavy and illustrates the fallacy of over-dependence on single safety margins. More particularly, it was a warning that not all species of even a single avian order respond in like fashion. I myself have found dead pheasants and doves with yellow-dyed grains in their crops. Following aerial distribution of colored 1080-treated grains over thousands of acres of California forest, bird loss was more clearly documented than were rodent losses (E. G. Hunt, personal communication). These qualifications notwithstanding, the approach is a good one. There is clearly a need for more work on this subject.

In summary, these conclusions apply to the three major groups of animals in which repellency is used as a method of pest control. Insect repellents are not useful on a large scale in the field. Mammal repellents, particularly as applied in reforestation, are somewhat hazardous, but different chemicals could reduce this hazard. Repellents to reduce bird depredations are generally ineffective, but colored grain baits for rodents provide some measure of protection to birds. In all cases the opportunities are great for further imaginative research. Many protective values can be served by an effective repellent.

10   SUBLETHAL EFFECTS

Control measures, whether or not conducted with toxic chemicals, have effects on surviving individuals. These effects may be slight or marked, and they may be echoed in noticeable changes in the behavior of populations. They may also affect succeeding generations.

It is curious that individual animals are commonly regarded as static units, responding uniformly to any stimulus irrespective of time or situation. The reason that animals for laboratory testing are reared rather than caught as wild animals is to ensure homogeneity of population or uniformity of response. Precise conclusions and predictability should be the result. The fact that research workers in nutrition and toxicology, for example, insist on large numbers of individuals per test and replications of tests is évidence that neither result is fully achievable. Yet there remains a widespread emphasis on the static individual as the measure of response. In pesticidal control tests the most notable response of the individual is death. Transferred to populations, the single, most widely employed measure is percentage of death among a known number. This measure is applied both in laboratories and in nature. For a majority of applied biologists, death measurement satisfies the economic or regulatory goals intended.

The good biologist understands the continuum of growth in the aging individual, the continuity of time and generation that is reproduction, the "steady state" that physiological reactions seek to establish. He knows the interacting relations among individuals of the same and other species. He knows that populations are not simply additive aggregates of the responses of the individual. And he knows that adaptation in organisms is the end result of all these and more dynamic processes.

With such awareness a biologist finds it inconceivable that changes short of death induced by control measures are not both common and important. This is a logical position, but one not considered in field data. Most of the data available do not reflect subtle effects. Strong dependence must therefore be made on information judged to apply to field situations, until such time as more direct data become available.

There is no doubt that individual and population characteristics change with changes in density. The basic cause is competition for food, shelter, and mates. The extensive studies of David E. Davis and colleagues on the rat populations of Baltimore clearly document the relations among population density, social interactions, and reproductive performance. Christian's general theory on population regulation in mammals is based on endocrine changes and "sociopsychological" phenomena (Christian, 1959). Implicit in the studies of Davis and Christian is that changes in population size—from whatever cause—produce profound changes in the physiological make-up and reproductive performance of individuals (Christian and Davis, 1957). Similar changes occur in insect populations during alterations in abundance (Wellington, 1960). The large-scale rearing of insects for mass release as biological-control agents has repeatedly produced effects of this kind.

The pertinent point here is that control measures causing reduction in numbers inevitably yield subtle sublethal effects on survivors, whether or not the survivors are affected by control chemicals themselves. It is a profoundly important reaction, and one grossly underemphasized.

The various sublethal effects of toxic chemicals can be studied separately; I propose to describe them here in that way, adding only that such a description is purely for convenience. The effects may be categorized under vigor, behavior, growth, and reproduction.

Vigor

A vigorous animal is healthy and active, able to perform its necessary functions satisfactorily. Vigor is a vague word and better described in its components. There has been voiced the suspicion that the total well-being—vigor—of desirable animals is undermined in areas under intensive chemical-control regimes. Clearly, animals with acute poisoning symptoms, even though they survive, are not vigorous; there is a reduction in the ability of such animals to perform in their natural manner. I do not know of a single field study that offers confirmatory data on population-wide depression of all normal functions. Perhaps the nearest to true data is a composite conclusion derived from studies of animal populations in the fire ant control area. In that instance, however, the emphasis was only on reduction in population numbers and interference with reproductive success. Questions still remain unanswered concerning effects on viability, selection of "resistant" types, altered physiological function, carcinogenic and mutagenic effects, and so on. Effects of this character have been shown to occur under laboratory conditions. Their applicability to animals in the field, particularly vertebrates, has yet to be shown.

Let me illustrate how they might apply. Chronic toxicity studies show that viability can be reduced by continued exposure to sublethal quantities of agricultural chemicals. Moreover, these effects may be passed on to offspring. DeWitt (1956), for example, demonstrated that sublethal amounts of chlorinated hydrocarbon insecticides in the diets of adult quail and pheasants lowered the viability of the chicks. I do not know of any demonstration that individuals initially surviving chemical treatment of natural populations have shorter lives. But it is likely, and I believe that it occurs.

Cell structure and behavior might well be altered by exposure to toxic chemicals. Mutagens—chemicals capable of producing genetic change—are common among hydrocarbons. There is abundant laboratory evidence that chemicals similar to kinds used in agriculture can be mutagenic. Geneticists have, in fact, used them to produce experimental changes in crop species. But I know of no data from nature that mutagenesis has been induced by toxic chemicals.

The same conclusions apply to carcinogens—chemicals that cause unregulated cell reproduction, i.e., cancer. There are a number of agricultural chemicals that are carcinogens under some conditions. An example is the common miticide, aramite. The cranberry "scare" of 1959 was based on residues of the herbicide aminotriazole, a chemical quite capable of producing cancer in laboratory rats at very high concentrations. Nonetheless, although it is possible, there is no evidence that agricultural chemicals are carcinogenic under field conditions.

Similarly, regarding increased susceptibility to disease, altered metabolic rates, and other effects that might be included under vigor, there is presumptive evidence from the laboratory that they might occur, but no conclusive evidence derived from nature. The general conclusion that the vigor of wild animals is unaffected by environmental contamination by toxic chemicals reflects ignorance, not knowledge. In the other categories—more specific components of vigor—the evidence is somewhat more reliable.

Behavior

Behavioral changes following intoxication are rarely studied in the field. It seems clear that surviving but affected insects frequently expose themselves to predation. I have observed this same phenomenon in fishes exposed to chemicals. Herons, egrets, and gulls congregate where fish are driven to the surface by the need for greater oxygen supply. During poisoning campaigns, field rodents appear on the surface, exposing themselves to predators (see, for example, J. J. Craighead, 1959)

The suspicion is strong that cholinesterase-inhibiting insecticides produce sublethal behavioral effects of a new kind. Shellhammer (1961) studied the relationship between cholinesterase levels in the brain and learning ability in wild mice. Psychologists had previously demonstrated that the ability of mice to learn was correlated with the level of brain cholinesterase. Shellhammer depressed cholinesterase levels 25 to 50 per cent in two genera of wild mice by injections of parathion. After cholinesterase levels had returned to normal (usually 3 or 4 weeks), he subjected the mice to a series of learning trials. In essence, he discovered that once-poisoned mice had "forgotten" previous learning trials, had to be "retaught" the tests, and, moreover, performed in these tests more poorly than unpoisoned animals or, for that matter, than they themselves had performed before being poisoned. A number of other behavioral changes occurred consistently. He had in effect produced "dull" mice, even though by usual standards of measurement they would have been pronounced normal. One wonders how spray operators, whose cholinesterase levels have on occasion been shown to be reduced by comparable percentages, might perform in test-retest situations.

Shellhammer's work may be more than academic. Two Australian scientists studied fourteen men and two women in whom mental aberrations appeared after prolonged exposure to organophosphorus insecticides (Gershon and Shaw, 1961). All persons were professional users of insecticides. Three were scientists, eight were greenhouse employees, and five were farmers. Of the sixteen, five had symptoms like those of schizophrenia with marked delusions. The others were persistently and severely depressed, and some had considerable memory loss, sleep-walking, and speech difficulties. When occupational exposure to pesticides was withdrawn, all individuals recovered. Nonetheless, clearly shown for the first time is the fact that pesticides can cause temporary mental aberrations.

Weiss (1961) recently established that cholinesterase-inhibiting insecticides profoundly depress enzyme levels in fish, requiring a month or more before return to normal levels. The pattern of response varied with species. Weiss' studies would form an excellent base from which to study behavioral changes resulting from sublethal poisoning.

In a prior section I alluded to Tevis' (1956) work describing the change in behavior when forest rodents are exposed to sublethal amounts of poisoned conifer seeds. After sublethal poisoning, the resident mouse population refused to eat any seed. Behavioral avoidance of this kind is called "bait shyness," and limits the efficacy of control techniques which reduce rodent numbers. In this instance, however, learned avoidance could be put to good use in reforestation. Unfortunately it has not been applied as widely as it should be.

Significant behavioral changes do occur in response to sublethal amounts of toxic chemicals. Their general occurrence in nature is not certain, but they are sufficiently well established in limited situations to warrant further study.

Effects on Growth

In laboratory studies of chronic toxicity one of the standard measures is rate of growth. A significant deviation from normal (usually a depression or slowing of normal rate) is interpreted as an important consequence of continued ingestion of toxic chemicals. Common as is this parameter in laboratory studies, I know of no study in the field in which this measure was a central focus. There have been repeated incidental observations in which was noted a change in the length of time between developmental stages in insects.

Biological reaction to pesticides may be direct or indirect. In recent years the importance of fertilizers and pesticides in altering the amounts and availability of basic nutrients (particularly nitrogen, potassium, and phosphorus) has been appreciated. The nutrient state of host plants affects growth phenomena in insects. Nutritional effects can be extended to changes in development rate, fecundity, and mortality.

Accelerated growth in mites indicates that soil insecticides may increase the nutritional value of host plants since it is known that mite growth rates respond to changing levels of potassium, phosphorus, and nitrogen contents of plants (Rodriguez, 1960). Some hormone herbicides applied to plants increase reproduction in feeding aphids; others increase mortality or reduce fecundity. C. A. Fleschner and other workers at the University of California's Riverside campus have shown that DDT has a stimulative effect on mite populations, in part nutritionally based. Rodriguez concluded that several kinds of herbivorous arthropods benefit from an oversupply of nitrogen and that the host plant can grow more profusely with the addition of insecticidal and fungicidal sprays or soil insecticides.

Growth-regulating substances are particularly effective as herbicides. They are not without effect on animals. Widespread outbreaks of aphids in New Brunswick grain fields were associated with selective herbicidal treatment with 2,4-D (Adams, 1960). This hormone reduces the numbers of predaceous coccinellid (ladybird) larvae and adults, the normal predators of aphids, by direct mortality of larvae, by producing larval deformity, and by delaying the onset of pupation.

Vernon Applegate, instrumental in the successful development of chemicals to reduce lamprey populations in the Great Lakes, noted in a public address that the chemicals his group uses also hasten the development of aquatic insects. If widespread, a growth stimulation of this kind could have serious ecological repercussions.

Indirect sublethal effects of general consequence are probably nutritional. An awareness of this importance is appearing in the literature.

Currently little studied are those more direct effects on animals which would seem to arise from chemicals of lower toxicity, particularly those depending on growth alterations for control purposes. Both kinds of effects have already precluded lasting benefit of pesticide treatments in some instances. A conservative prediction is that sublethal effects will become increasingly obvious, convincing, I would hope, even doubting applied biologists that the responsibility for control methods does not stop with the initial recommendations.

Reproduction

In a general sense almost all crop-growing practices influence the ability of adult animals to reproduce and of young animals to survive. The character of the agricultural environment and of the animal's ability to adapt to it are basic matters, but not the subject here. If adaptation to a crop environment is assumed, in what manner do control chemicals affect the reproductive success of an animal population?

Pesticides may either enhance or inhibit reproductive success. Entomologists have contributed a number of observations describing stimulatory effects on insect and mite reproduction. The effect has been noted repeatedly in mite populations exposed to DDT. In a species that does not react directly to the toxic nature of a pesticide, the cause of increased fecundity is basically nutritional. At low levels, organic pesticides and fertilizers provide a supplementary source of nitrogen, phosphorus, and other nutritive elements. The usual pathway by which these essential chemicals reach the insects is through the host plant. The "vigor" of a plant is reflected in the well-being of the arthropods feeding on it. More specifically, the amounts of several key nutrients in plants have been shown to vary widely with the chemical regime to which plants have been subjected (Fleschner, 1952, for example). The reproductive ability of dependent insect populations has been shown to be correlated with the chemical constitution of host plants. The significance of such responses, although clearly established in insects, is conjectural in vertebrates. The sole example of response in vertebrates to changes in plant nutrition shows only indirect effects on reproduction. This is nitrate poisoning among domestic herbivores occasioned by alterations in nitrate and nitrite levels in plants through the effects of herbicides and fertilizers. Neither a stimulatory nor a depressive effect on reproduction from altered nutrition has been demonstrated in wild species other than arthropods.

Direct toxic effects on adult animals, in turn affecting reproductive capacity, are a different matter. The continued ingestion of sublethal amounts of pesticides has clearly led to depressed reproductive performance in vertebrates. Stored residues of chlorinated hydrocarbon insecticides tend to concentrate in reproductive organs (Genelly and Rudd, 1956). Among wild species this unusual concentration has been noted in gallinaceous game birds, in fish-eating birds, and in trout. Fishes and birds in particular demonstrate insecticidal accumulation in their eggs (see, for example, Moore, 1962). The best evidence for reproductive suppression comes from controlled experiments with game birds, but there is good evidence from the field that parallel effects occur or are likely to occur.

The basic measures of reproductive success in birds are egg production, fertility and hatchability of eggs, and viability of young. R. E. Genelly and I (1956) conducted tests under controlled conditions to determine the alterations in these components in ring-necked pheasants as a result of sublethal ingestion of three commonly used insecticides. There were two test groups, each fed a different level of chemical, and one control group. Reproduction was affected in all chemically contaminated birds, in some only to a minor degree. Food consumption and egg production were strongly correlated; not all contaminated food was acceptable to the birds, which thus consumed less. Egg production was significantly depressed by toxaphene or dieldrin. Lowered fertility and hatchability in some groups were related both to insecticidal concentrations in eggs and to poor condition of adult females. Initial mortality of young (of course, only that caused by insecticide content of eggs was considered) was greater in all chemically contaminated groups than in control groups. Viability of young was affected most by dieldrin and DDT. In composite terms, reproductive success was 70 per cent in control birds, and about half that figure in treated birds when all chemicals were compared. Recall that all doses were calculated to be below lethal levels for adults. The observed effects therefore had to come from storage and accumulation of insecticides in tissues of adult birds and in eggs. DeWitt (1956) showed similar responses in quail and pheasants. An interesting observation was that the fertility of eggs was slightly enhanced at low levels of DDT contamination. A parallel condition has been noted in the fertility of insect eggs.

Field evidence supporting the charge of suppression of reproduction in vertebrates by pesticidal chemicals is chiefly suggestive. Growth, survival, and reproduction in wild species represent a biological continuum in animals that usually live not more than a year or two. Perhaps the best method of limiting the subject for discussion is to define "reproductive effects" quite narrowly. So limited, effects on reproduction would be confined to pesticidal effects that are sublethal to adults but may not be for young, and that result in a decreased overall reproductive success in a given season. Excluded from consideration by our definition would be mortality to adults during the nesting season that results in nest failures and starvation of young (Hunt, 1960; Wallace *et al.*, 1961). Also excluded would be direct mortality of young, as would be the continuing low numbers of population following heavy mortality. All are in a broad sense reproductive failures, but would only complicate discussion here. The phenomena of food-chain relationships and of greatly delayed effects from slow accumulation of residues in tissue are reserved for later detailed presentation (Chapter 20).

Given the above restricted definition, one must concede that evidence of reproductive suppression in wild vertebrates is largely lacking. Available evidence is indirect, fragmentary, and not always logically applied (*post hoc, ergo propter hoc*). Let me list a few situations in which reproductive effects are probable.

For twenty-five years A. H. Miller, one of the world's leading ornithologists, kept detailed field notes on birds at his resort cabin on the shores of Clear Lake, California. He could not have known how important his data would become in elucidating the complex sequence of chemical transferral at Clear Lake. Miller was the first to note the lack of reproductive success in the local breeding population of western grebes in the year following DDD application to the lake (Miller, field notes). In subsequent years large numbers of adult grebes in breeding condition died. Only recently, three years after the chemical applications ended, has breeding been again successful among surviving adults. There is no doubt in my mind that breeding in this grebe colony was interrupted by initially sublethal effects of ingested DDD and its metabolites.

A similar sequence probably explains in part the population declines in bobwhite quail and other birds in parts of Alabama heavily treated with heptachlor and dieldrin for fire ant control. Local populations of quail declined in number beyond that for which adult mortality alone could account. Both wild and domestic turkeys very nearly failed to reproduce at all. At one site, three hens laid fifty eggs of which seven hatched; all seven died soon after hatching. Turkey populations on nearby untreated areas reproduced normally. Throughout the area ground-dwelling birds suffered highest mortality; it is probable that reproduction in birds of this stratum would also be affected most heavily (DeWitt and George, 1960).

Chemical suppression of reproductive efficiency has been suggested to explain lowered numbers of young in woodcock. In this instance the exposure was 1 pound of DDT per acre over New Brunswick forests (Wright, 1960). It is also possible that woodcock on wintering sites acquired tissue residues of insecticides that they carried back to breeding grounds in the spring. The hypothesis gains credence when one recalls that a good part of the woodcock in northeastern North America winter in Louisiana, large sections of which have been heavily treated with heptachlor.

There are other similar situations in which sublethal effects on reproduction are likely—in Dutch elm disease, Japanese beetle, and white-fringed beetle control areas, and in areas in which seed treatment is common. R. E. Genelly and I noted (Rudd and Genelly, 1955), for example, mortality in pheasants in rice-growing areas where heavy concentrations of DDT (20,000 parts per million) were placed on seed rice.

The likelihood that heavy concentrations of pesticides in rice-growing areas will produce profound but subtle effects on reproduction has been enhanced in 1961 by pilot investigations of the California Department of Fish and Game (E. G. Hunt, unpublished). These studies— which continue—center on ring-necked pheasants near Gridley, California, where there are contrasting areas: one in which pesticides are commonly used in rice and other crop production, and an adjoining wildlife refuge where pesticides are not used. It is important to realize that in this investigation pesticide residues and derived effects are consequences of completely normal agricultural practices and that, interrupted only by the need to sample wild populations, pheasants are living under conditions normal to the area. Hens were shot at the nest and the clutch of eggs removed for continued incubation at a game farm or for chemical analysis. Chemical residue checks of fatty and muscle tissues were made on 50 hens, 29 from the treated area and 21 from the nearby refuge or "control" area. Every hen contained DDT or its metabolite DDE. The amounts ranged from 19 to 2930 p.p.m. (averaging 741.05 p.p.m.) from the treated area and from 0.4 to 7.2 p.p.m. (averaging 2.14 p.p.m.) from the control area. Although respective values from the two areas were greatly disparate, contamination was general. Moreover, other pesticide residues were present. Dieldrin was present in 37 samples in amounts to 48 p.p.m. Five other types of pesticide residues were also present in tissue to some degree. Egg yolks derived from normal clutches contained 1.9–406 p.p.m. DDT and 0–1.31 p.p.m. dieldrin. Clutch size and fertility did not differ between the two areas. But the mortality of young in the first four weeks following hatching was 46.6 per cent in the treated area and only 27.0 per cent in the control. The per cent of defective young from eggs in the study area was 25.0 per cent, nearly twice that of the control area (12.9 per cent). The increase in loss from deformities and post-hatching inviability in pheasants in this area where heavy use of insecticides is common practice is probably caused by transfer of residues from hen to young via the egg. More intensive work now underway will further clarify sublethal effects, formerly only suspected, from completely normal pesticidal practices.

In summary, population declines may be brought about by an increase in mortality rate (abundantly shown) or by a decline in the birth rate (poorly shown). A number of dead animals found in a treated area may indicate an increase in mortality, but the signs of a decrease in birth rate are far less obvious. Despite the lack of clear, abundant evidence for reproductive suppression under field conditions, both laboratory and field studies strongly suggest that these effects occur. Further careful investigations are certainly warranted.

11 · RESISTANCE TO INSECTICIDES

No problem in pest and vector control is more important than that posed by the development of resistance to insecticides. More than 120 of the some 5000 species of insect and mite pests on earth are now to some degree less susceptible to control chemicals than they were fifteen years ago. This seemingly small fraction not only is increasing to include more species, but also it unfortunately represents most of the major insect pests and disease carriers. The phenomenon is not restricted. It occurs throughout the world in urban, crop, farm, and marsh environments. The Utopia envisaged two decades ago, with freedom from insect pests assured by DDT, began to fade as insect resistance appeared. Neither legislative nor chemical action could long confine the astonishing adaptability of insects. A purely biological phenomenon has done more to encourage a return to basic biology in pest control practices than have all prior critics of chemical uses, whatever their arguments.

At its simplest, resistance means that insects can no longer be killed by levels of chemical exposure that were once successful. In practice it means that more chemical must be applied more frequently to achieve the same control; if, for reasons of hazard or cost, higher rates cannot be used, other chemicals or other methods of control must be substituted. Alternative methods are not always immediately available. The problem can therefore become more serious than it was before resistance developed. To the research entomologist, resistance means that the segment of a population able to withstand exposure to toxic chemicals has enlarged. When this segment becomes a prominent fraction of a population and resistance is continued in subsequent generations, whether or not further exposed to chemicals, a resistant strain is formed.

The widely accepted meanings of resistance given above, whether defined from the point of view of either the applied or the research entomologist, do not give the causes. They are simply descriptions of terminal effects, and do not give the necessary emphasis on the biological basis for resistance.

Biologically speaking, the ability of insect populations to become relatively resistant to insecticides should be understood as only a special case of the far more general phenomenon of adaptability of populations to environmental changes. This adaptability is a consequence of the genetic variability of all sexually reproducing animals. No species, race, or strain embodies only a certain type, a "norm." An individual organism never represents such a norm; it is rather the carrier of a genetic constitution unlike that found in any other individual. The processes of genetic segregation and recombination inevitably result in new genetic constitutions in each succeeding generation. If an environment, whether natural or artificial, favors a certain group of genes—a particular genetic constitution—the favored group increases and tends to replace the unfavored. Insecticides are, then, often merely agents that eliminate the more susceptible individuals, which may encourage the increase of more resistant ones. Mutations are necessary neither to this process nor for continuing change to be the rule. The great number of recombining genetic possibilities provide sufficient variation to maintain change. This sequence is not the exception. It is the raw process by which virtually all evolutionary change is achieved. In the development of insecticide "resistance" we have had an opportunity to observe evolutionary change in an amazingly short period. Even what is termed rapid evolution requires time spans beyond a single human life expectancy. Only the ubiquitous distribution of insecticides can explain their "strength" as intense selective agents responsible for world-wide change in the genetic make-up of entire species populations in so short a time.

It should be emphasized that the term resistance as normally used is applied only to the survivors of an exposed population. The rest of the population has been *killed.* I must emphasize the word because too many confuse resistance with tolerance. Resistance does not apply to the ability of an organism to withstand exposure to increasing amounts of chemicals during its lifetime. This is individual tolerance, and it varies with individual, chemical, physiological state, and so on. In most cases, individual tolerance levels are much below those in which resistant forms can survive. I have heard the question: If insects can become resistant, why can't birds, human beings, or other higher forms become resistant too? They might, over sufficient generations subjected to like stresses. It should now be clear, however, that none of us wants "resistance" under the conditions necessary to achieve it.

Although resistance is a consequence of a normal biological process, the identification of resistant forms has not centered on general biological occurrence. Understandably, the resistant arthropods identified thus far are all forms of some immediate importance to man. Resistance probably occurs in nature in arthropods not so directly related to man's efforts to safeguard himself and his crops. Of the more than 120 species of insects and arachnids that are known to have developed resistance, about half are agricultural pests and the other half are arthropods of public-health or veterinary importance (A. W. A. Brown, 1958). Identified by taxonomic groupings, 54 are Diptera, 23 Hemiptera, 16 Acarina, 14 Lepidoptera, 6 Coleoptera, and the remaining 9 are thrips, fleas, cockroaches, and lice. It is clear that resistance is broadly represented among pest invertebrates. The numbers of affected populations and of species are continuing to increase.

The occurrence of resistance is related to the widespread use of persistent insecticides in the twentieth century. It was first observed over 50 years ago, but only in the post-World War II period has it become so widespread. The history of the spread of resistance is as follows (A. W. A. Brown, 1958):

1908   Lime-sulfur on San Jose scale in apples
1916   Hydrogen cyanide on red and black scales in citrus
1928   Arsenic on codling moth in apples

1935 Arsenic on ticks of cattle
1938 Hydrogen cyanide on citricola scale in citrus
1939 Hydrogen cyanide on flour beetles in grain
1942 Tartar emetic on thrips in citrus
      Phenothiazine on screw-worms of cattle
1946 DDT on houseflies
1947 DDT on mosquitoes
1949 Parathion on greenhouse mites
1951 DDT on body lice
      DDT on codling moth in apples
1952 DDT on mosquitoes (behavioristic resistance)
      Pyrethrum on houseflies
1955 Parathion on houseflies
1963 Many insecticides and over 120 species of arthropods, and the recent
      suggestion (1962) that vertebrates with brief generation spans too
      may become resistant to insecticides.

This list identifies only the first occurrences of resistance to major insecticides. Many species and environmental situations are not included, nor are the complexities of cross-resistance shown. More than 30 species of major pests are resistant to DDT alone, and most of these have some degree of resistance to related organic insecticides, whether or not previously exposed to them. The acquisition of resistance to related insecticides in the process of achieving resistance to the primary exposure is common. Such cross-resistance falls within three groups. One includes DDT and its analogues, methoxychlor and DDD. The second is dieldrin and the related chemicals, aldrin, endrin, heptachlor, chlordane, toxaphene, and BHC. The third type of resistance is to organophosphorus compounds, which can in turn be broken into subtypes.

Resistance may not be maintained at the same level if chemical exposure is withdrawn. Normally this is caused by the diluting effects of matings with "wild type" individuals from surrounding populations in which resistance has not developed. Although entire populations may show decreased incidence of resistance, only rarely do once-resistant insects return to original levels of susceptibility. Moreover, from experiences in many countries with many insecticides it is clear that resistance is re-established very quickly. In other terms, initial resistance may require six to twenty generations of insects for first establishment, but its re-establishment upon second exposure will require only three or four. The second exposure need not be to the same chemical (Natl. Research Council, 1952).

The specific factor which favors development of resistance is rarely known, though a great deal of research work has been done to help clarify the causes. The individual characteristics for three general qualities influence insecticide resistance: physiological, morphological, and behavioristic (Briejer, 1958). The first quality pertains specifically to the biochemical capacity to render a poison harmless, a detoxification similar to that performed (within limits) by the liver in vertebrates. The second quality involves a physical attribute, possessed normally by some individuals in a population, serving to prevent ingress of a chemical. A thickened cuticle is one example. The last quality may or may not include characteristics which lead to resistance: It is an avoidance of contact with an insecticide by some behavioral trait. Some mosquitoes, for example, will not alight on treated surfaces, or they might select new breeding sites or elect not to enter human habitation. If the trait has been selected for by insecticidal treatment that has removed a large segment of the population not showing this behavior, the quality is as assuredly "resistant" as any physiological or morphological manifestation. The behavior of an individual is, after all, the functional expression of all its component parts. If environmental interactions of a new behavior pattern are emphasized, then we may properly speak of "ecological resistance" as an extension of the newly acquired behavior pattern.

In one sense it is possible to classify all resistance problems under an ecological heading. Although the elucidation of precise mechanisms normally comes from the laboratory, the significance of resistance lies in nature. This significance is twofold—an illustration of the amazing adaptability of insects, enabling rapid evolutionary change, and, practically speaking, a reminder that food production and disease prevention are at base complex biological processes neither fully understood nor controllable. We are not yet freed of our dependence on nature or of the challenges it provides.

There are two corollary questions that are important enough to be posed if not answered. Are insect forms other than pests becoming resistant? And does an insecticide-resistant population have the same density (numbers per area unit) as an unresistant one?

Regarding the first question, several species of laboratory insects have, of course, been subjects of induced resistance. There has been a little effort to select for resistance in beneficial predatory insects prior to their release as biological control agents; it is possible to do so, but not practical. There is only the suggestion that crop insects other than pests can become resistant. The possibility has not been investigated either for its occurrence or for its exploitation. In short, the question referring to arthropods can be raised but not answered.

Very recently, however, a group at Mississippi State University has offered strong evidence favoring the likelihood of insecticide resistance in vertebrates (Boyd *et al.*, 1962, 1963; Vinson *et al.*, 1962). Studies there centered on cricket frogs (*Acris*) and mosquito fish (*Gambusia*) living in ditches, ponds, and bayous adjacent to heavily treated cotton fields. DDT has had a long history of use on these fields although toxaphene, methyl parathion, and endrin are also commonly applied. Twelve aerial applications of pesticides per growing season are not uncommon and high toxicity from drifting sprays has been attested by occasional mass mortality of vertebrates ranging from fishes to water birds. Survival of cricket frogs and mosquito fish in such an environment led to experimental testing with DDT, clearly confirming that these organisms indeed had an unusual capacity to survive in contaminated waters lethal to individuals without prior exposure. Moreover, cross-resistance to aldrin could also be shown even though aldrin had not been applied to nearby cotton fields. Presumably, intensive selection for genetically adaptable fractions of total populations occurs among these fishes and frogs in ways completely comparable to those among arthropods.

The answer to the second question is "no." It should be added, however, that the reasons for greater abundance of resistant insects are not always clear. And, further, that in a few situations quite the reverse is true. A population increase in resistant forms may occur because of more successful reproductive performance (related to greater vigor), or it may occur because of release from natural controlling factors. While many earlier studies showed little or no differences in fecundity between resistant and unresistant strains, more recent work shows that egglaying is greater in some resistant strains (Gratz, 1959). The earlier studies dwelled on DDT resistance. Dieldrin resistance particularly seems to be responsible for greater reproductive success. The matter is not settled. In general, it would seem that most resistant strains have no particular added vigor (reproductive potential as one expression of it) but that variation occurs ranging from poorer to greatly enhanced reproductive success. The many observed increases in population numbers of resistant strains are probably not due to enhanced vigor or reproductive performance.

The point should be made that the effects of insecticide treatment range well beyond the target species. Accordingly, greater numbers of a resistant insect population must derive not so much from an inherent advantage but from insecticide effects on nontarget species normally competing with or preying upon the resistant form. Release from natural limitations of numbers has been repeatedly shown in insects of agricultural and medical importance (see Chapter 21). Let me reinforce this with a few examples from public-health programs (Gratz, 1959). In Zanzibar, bedbug, flea, and chicken-mite populations increased following dieldrin spraying. In Southern Rhodesia, bedbugs resistant to BHC, DDT, and dieldrin increased to numbers not previously known in the area. In Phoenix, Arizona, resistant houseflies emerged in much greater numbers from dieldrin-treated privies; the reason seems to be the elimination of a highly susceptible competing species of fly whose presence normally renders privy contents unsuitable breeding places for houseflies. A. W. A. Brown (1958) calls attention to increased housefly production in Liberia, Nepal, Saudi Arabia, Japan, Sicily, Sardinia, Egypt, Kenya, and Tanganyika following spraying. So the list grows; the problem is world-wide.

I do not know how far effects of the kinds described above can be extended to so-called natural populations. Moreover, the processes establishing insecticide resistance and the resultant ecological consequences are consequences of biological phenomena. Where exposure to insecticides occurs, profound effects in nature can be predicted.

The occurrence and importance of insecticidal resistance in insects can be summarized as follows (see also Fay, 1959):

The numbers of resistant species and their areal distribution will continue to increase.

Resistance results from continued exposure of insects to toxic chemicals; it occurs in a wide variety of insects and is induced by a wide variety of chemicals.

Resistance is associated most frequently with persistent chlorinated hydrocarbon insecticides; where nonpersistent chemicals yield resistance, it derives from frequent applications.

Resistance to one insecticide frequently confers some measure of resistance to other related insecticides, even if there has been no exposure to them.

The resistance phenomenon is the most important biological limitation to successful insect control

Emphasis on persistent insecticides for pest control is slowly being withdrawn, and a partial return to basic sanitation, cultural practices, or biological control is under way.

Resistant insects generally do not have any greater vigor than unresistant ones, but their numbers per area unit are frequently greater. This is associated with adverse effects of treatment on competing or predatory forms normally limiting numbers.

There is no knowledge of the development in nature of resistance in nonpest insect species, and only recently has resistance been identified in fishes and amphibians.

Far more facts from the field are required to assess the full significance of resistance phenomena. Nonetheless, insecticide resistance as presently revealed presents a dismal forecast for the future use of chemicals in large-scale insect control programs.

12 · HUMAN HAZARDS

The aim of pest control is to benefit man—to protect something of value to him. Man devises the control schemes and manufactures and distributes the chemicals necessary in these schemes. He must also protect himself from harm from the chemicals he uses in pest control. The values he maintains and the desire for self-protection put him in a quite different category from the other organisms described throughout this book. He therefore justifies a special treatment as instigator of the practices that both benefit and harm him.

I do not intend to discuss pesticide hazards to human beings at great length (see also Chapter 10). Our self-interest is reflected in thousands of papers pertinent to human hazards. Some excellent reviews have appeared, and these should be consulted. I have referred particularly to those of J. M. Barnes (1958) and of W. J. Hayes (1959, 1960), which are characterized by great scope and excellent technical familiarity with public-health matters. I should add, however, that, where judgments of values are concerned, the views of these or other reviewers are as vulnerable to debate as anyone's views. The word "expert" must be reserved for technical mastery.

Human beings are better protected from pesticide damage than are other living things. The instruments of legislation, education, and morality assure that. In a statistical sense man is reasonably well protected from the direct effects of his own inventions. The doubts about safety, repeatedly raised, are of two kinds: concern over that fraction of individuals who accidentally come into direct contact with toxic chemicals; concern over the consuming majority who continually ingest low levels of insecticide contamination in their foodstuffs. For the moment I am exempting from discussion the small fraction of individuals who argue for no chemical control of pests whatsover.

Direct exposure to toxic levels of chemicals is chiefly accidental. Only in rare instances is exposure intentional. A number of suicides (and a few homicides) have chosen agricultural chemicals to accomplish their ends. Some of these chemicals are eminently suitable. Parathion, for example, has been used many times for suicide in Germany. The greater number of illnesses or deaths are not intended, and arise chiefly from carelessness or ignorance.

In households, adults may unknowingly ingest contaminated food or children gain access to stored pesticides or to the empty containers in which they came. Parathion and thallium, as examples, have both caused death from their presence in flour. An entire family became ill after consuming spinach brought in a jute bag formerly containing parathion; the spinach itself had been free of insecticides. Accidental death of children has been reported many times. Particularly responsible are rodenticides and organic phosphate insecticides stored in easily opened or breakable containers. One child, for example, died after drinking from a Coca-Cola bottle in which parathion had been stored. A girl died after breaking and spilling part of the contents of a bottle of TEPP on herself. Another died after playing in a large empty container that had held parathion dust. Gross carelessness and ignorance are as much the causes as the pesticides themselves. Exposure of this character continues to occur, and can perhaps always be expected. In the United States about 150 deaths a year are attributed to pesticides. In one year (1956) almost half were due to the "older" pesticides, arsenic and phosphorus. In that year only 35 of the total number were linked with the newer synthetic compounds. The Public Health Service agencies and poison information centers in the United States do all possible to regulate and educate against hazards of this kind. The World Health Organization, through its Division of Environmental Sanitation, assembles published instances of illness or death from pesticides throughout the world. From these reports it is clear that the problem has no national boundaries, but that the extent to which a people appreciates the toxicity of pesticides correlates with the incidence of poisoning.

Occupational exposure is much more likely to produce illness or death from pesticides than is accidental exposure. Each year hundreds of cases are recorded throughout the world. Most instances involve appliers who through carelessness or ignorance fail to heed or understand precautionary recommendations. Men working in the manufacture or formulation of pesticides are trained to avoid hazard; accidents are therefore seldom in these exposures.

Accidents among spray operators have mounted as the use, variety, and distribution of pesticides have widened. Tempting as it might be, the increase in accidents cannot be laid on the doorsteps of illiterate native peoples. California, for example, where pesticides are used in greater volume and variety and where their use is more closely regulated and observed than in any comparable area on earth, has experienced a striking increase in poisoning incidents. Almost 500 cases of poisoning with farm chemicals were reported in 1957. The greater number were caused by exposure to organic phosphate insecticides. Only one instance in that year was fatal—a "swamper" died within five hours of washing out an airplane previously used to apply demeton (Kleinman *et al.*, 1960). In more recent years the rate of poisoning has increased. Occupational poisoning from agricultural chemicals in California more than doubled in the years 1957 through 1959 (Calif. Dept. Publ. Health, 1961). Almost 1100 cases were reported in 1959. Of those, almost half were systemic poisoning, by far the greater percentage (81.7 per cent) caused by organophosphorus insecticides. Particularly vulnerable were Mexican *braceros*, seasonally imported to work in California fields. On several occasions entire crews of men became ill from residues of parathion, Thimet, and phosdrin. Poisoning of this kind must be checked in large part through education and care in handling chemicals. Instruction, however, is not enough; much more needs to be learned about the decomposition and manner of exposure of applied chemicals before correction can be complete.

The agricultural aircraft industry is particularly subject to pesticide injuries. Poisoning is in fact the largest single hazard of the industry, and its importance is rising with each passing year.

A certain fraction of injury can be expected where pesticide use is common. Ignorance is at the root of it; theoretically at least, injury rates can be reduced through education. Reduction is less likely with illiterate peoples whose funds are inadequate for proper application of chemicals and adequate safeguards. A Mexican peon who cannot read cannot be protected by labeling precautions.

In many countries the conditions under which pesticides are applied are primitive, and great risks in pesticide applications are inherent and assumed. At the same time, in those countries the need for heroic measures in pest control is abundantly clear. The antimalarial campaigns, by themselves, have succeeded in relieving human misery to a degree unprecedented in history. Whatever qualifications one might make regarding the hazards of pest control, one must concede that the risk incurred must relate to the need. In those countries the need is great; it compares in no way with the indulgent requirements of an affluent people as found in the United States.

The question of the significance of residues in foodstuffs has brought forth a national controversy in which large segments of the public and several state and federal agencies have found themselves at odds. The argument rests on several conflicting elements:

Some individuals, for convenience grouped as "organic farmers," want no use of agricultural chemicals; their emphasis is on natural control.

Some chemical and agency representatives have overzealously promoted pesticide use.

Excessive amounts of chemical residues have on occasion been found on foodstuffs.

Pesticides have on many occasions killed bees, domestic animals, and wildlife.

Governmental control campaigns are believed by many to be too single-minded; insufficient emphasis has been accorded to the total effects of pesticide use (for example, in the fire ant control campaign).

The rapid increase in the use of toxic chemicals, coupled with their known potential for harm, has not been paralleled by the equivalent increase in hazard research and inspection activities.

An atmosphere of apprehension has been created among many people by "scare" publicity.

Some unanticipated responses to chemicals have further increased apprehension (for example, accumulated contamination by several stable chemicals in tissues; rises in levels of residues in soil and streams; in short, a wider exposure to residues than expected).

.Not all the foregoing can be considered directly pertinent to the question of harm to human beings from residual chemicals. Yet in all major investigations these subjects are intertwined. They were present in the Delaney Committee and later congressional hearings, the legal action against the gypsy moth control campaign on Long Island (the so-called DDT trial), the two special governor's committees (Wisconsin and California), the investigations in the fire ant control campaign, and the deliberations of the National Research Council (see Chapter 5).

Not all of these ingredients of controversy can be weighed equally, nor can all be reduced to factual appraisal. As a means of making a general appraisal, let me pose what I consider to be four fundamental questions, and reply with short opinions of present status.

Do foodstuffs become contaminated with residual pesticides? Yes, they do, but occurrence depends on the type of foodstuff, its manner of production, and to some extent the region of the country from which it came. The amounts are usually at trace levels, well below residue tolerance levels set by the Food and Drug Administration. Excessive levels—that is, those exceeding legal tolerance limits—occur rarely, and are normally the result of failure to follow precise recommendations for application. The only important exceptions occur in milk and in the fatty tissues of meat animals. By law no residues can be present in milk; yet sampling reveals consistent contamination of marketed milk, usually in trace amounts (see Chapter 13). The contaminants are usually chlorinated hydrocarbon insecticides, most of which are not now recommended for use around dairy barns. Surviving residues from former treatments, the almost universal presence of DDT, the feeding of contaminated forage, or violations of use must be held responsible. Contamination of meats arises mainly from feeding contaminated forage. Probably more often than not, residue on forage is not suspected. Residual contamination of foodstuffs can be readily transferred to human tissues. In one study (Laug *et al.*, 1950) the average of DDT in the milk of twenty wet nurses in a Washington, D.C., hospital was found to be 0.13 part per million; none of the women was occupationally exposed to DDT. The average concentration of DDT in human fatty tissue has risen steadily over the last 15 years; in 1963 the figure stood at 12 p.p.m. Those with occupational exposure to DDT tend to have higher concentrations. The ubiquity of DDT residues in foods is the reason. In one random analysis of typical restaurant meals, all contained DDT. The average meal contained 0.2 p.p.m. (Walker *et al.*, 1954). In short, most meals and most human beings contain pesticide residues.

Are residues in foodstuffs a significant health hazard to human beings? The answer is perhaps debatable. Most reviewing authorities say "no!" A minority argue not only that present residues in human tissue are damaging, but that their presence may be linked to certain pathological conditions. Hayes (1959) contends that no important pathological changes can be associated with DDT residues in human tissue, and that insecticides, properly used, are not the cause of any disease either of man or of animals. He places the amounts of residue demonstrated to be present in typical meals at about one one-hundredth of the amount necessary to produce questionable liver changes in rats. Dr. Hayes fed human volunteers relatively great amounts of DDT daily for over a year without manifest pathological effects.

The amount of DDT and its metabolites in the fat of persons without occupational exposure in the United States has risen steadily in the past fifteen years. Hayes (1959) estimated the average amount to be between 5 and 6 p.p.m. in 1958. In 1963 the amount stood at about 12 p.p.m. Recent studies in England and Germany show an average concentration of 2 p.p.m. Dieldrin has recently been found in the fat of residents of southern England. It is quite likely that dieldrin as well as other chlorinated hydrocarbon residues will be found to occur widely among Americans when appropriate samplings are made (Kennedy, 1963).

My personal judgment is that present allowable residues in foodstuffs have not demonstrably caused ill effects of general distribution. Some individuals may be particularly sensitive to these residues, but establishing any link between these sensitivities and pesticides has been most difficult to do. Similarly greater susceptibility to disease has not been demonstrated. Nonetheless, hazard from ingestion of residues over very long periods is yet to be assessed. Should residues increase generally, any hazard would be magnified. Moreover, there is now no escape from these residues for most of the population. It would be somewhat more than embarrassing for our "experts" to learn that significant effects do occur in the long term. One hundred and eighty million human guinea pigs would have paid a high price for their trust. A serious moral question has yet to be answered. Have we the right to insist on free choice between residue-contaminated and uncontaminated foods? The current "residue" legislation in effect says "no!"

Are present governmental agencies able to monitor residue levels properly and to guarantee continuing conformity to safety standards? Not adequately, I think. I indicated earlier that "use experience" is strongly depended upon by manufacturers and regulatory and recommending bodies. Implicit in this dependence is the ability to detect violations or new patterns of residue distribution. The inspecting agencies are poorly manned for these tasks. There is, moreover, a tendency to treat low-level contamination rather lightly if enforcement would be difficult. This applies particularly to the producers of milk and other animal products. Countering this tendency is a general tightening of recommendations for chemical-application procedures, stemming largely from possible condemnation of products and from public censure. This is notably true in the agricultural experiment stations within the last three or four years. They are held legally liable should a recommended procedure result in residues exceeding established limits. Nonetheless, although conservative forces have recently come into play, an increase in residue contamination calls for much more inspection. A doubling of inspection personnel and residue research workers would seem justified at present.

Is a judgment of no present hazard to human beings from residual chemicals applicable to other forms of life? The answer is definitely "no!" Production and marketing of foodstuffs are well-understood processes. "Residue management," to coin a term, is currently inadequate, but, at least in time, some measure of understanding of levels in foods and their effects on human beings will be forthcoming. But there is only a meager understanding of the cumulative effects of residues in soil, of the amounts and the avenues by which chemicals escape from treated areas, of the amounts present in living organisms and the pathways by which these organisms concentrate and transfer residual chemicals, of the chemical transformations undergone in soil and in living things. It seemed to me curiously paradoxical that the Governor's (California) Special Committee on Public Policy Regarding Agricultural Chemicals (Mrak, 1960) could, on the one hand, conclude that no unmanageable residue problems existed and, on the other, recommend eleven points calling for more stringent administrative controls or stepped-up research efforts. With one of the committee's recommendations—a proposed Ecology Research Center—I can most heartily agree. It is, however, tacit recognition that "environmental sanitation" —a large part of which is concerned with residue chemicals—is an area requiring a great deal of further study.

IV · THE RESPONSES OF ENVIRONMENT TO CHEMICAL CONTROL

13 · PESTICIDE RESIDUES

The chemicals used to control agricultural insects are normally directed to the leaf surfaces of plants. A good portion, of course, misses the plant and is deposited on the surface of the ground. More is added to the ground surface by runoff from the plant. Still more is added from some pesticides applied directly to the soil. Any surface spray application will therefore contact both plant and soil surface. Rainfall or normal cultivation carries some of the applied pesticidal material into the soil itself. Without regarding natural dynamics, we can expect pesticide contamination to occur on the plant itself, on the surface of the ground, and in the top few inches of soil. Except for deeper-rooted trees and shrubs, the entire plant structure in a treated area—from root tip to meristem—falls within the contaminated zone.

The chemical mixture falling upon the plant or ground forms a *deposit* whose presence is necessary for pest control. This deposit does not normally retain its original character for long. Change in the deposit comes about as it is acted on by living systems and by the physical effects of heat, light, and water. The remainder of the transformed deposit is that to which I have repeatedly referred as a *residue*. The residue itself may contain reduced portions of the original toxic ingredient, metabolic derivatives of this chemical, physically transformed derivatives of quite different chemical structure, and surviving portions of the solvent and diluent *carriers* of the original material. The very wide differences in chemical responses to the even wider variables of nature preclude any precise definition of the word "residue." As will become clear, the chemical nature of residues, their impingement on living systems, and the particular meaning they have in relation to our subject intertwine to create an extremely difficult set of problems. Next to insecticide resistance, no problem of chemical pest control today has more significance.

All pesticide compounds in common use produce residues that survive for noticeable periods. The persistence of some may be for only a day or two; others, depending on soil, plant, and weather conditions, can survive ten to fifteen years or more. Both the value and the hazard of pesticide chemicals lie with that fraction of the residue that is toxic. With a chemical such as TEPP the toxic fraction may survive for only a day or two. Its residual value as an insecticide is therefore limited to this period. The control of hazard for this time depends on preventing access to the residue. For a short-lived chemical, essentially total success in preventing contact is feasible; beyond a few days—perhaps to three weeks—control of hazard is less successful. Toxic residues surviving more than a few days can create serious difficulties. There is no current indication that the problem of strongly persistent residues can be adequately resolved.

In most instances residues on foliage or on the immediate target do not have pesticide value for more than two or three weeks. Exceptions to this are in two categories: the "systemic" chemicals, which depend on delayed action, either chemically or physiologically, for their pesticidal value; and those most stable chemicals that lose their primary structure only slowly and are accordingly pesticidally active for long periods. In the first class are the hormones, herbicides, systemic insecticides, and, by easy latitude of definition, the anticoagulant rodenticides. The first two depend on translocation of the original material to a new site within the plant, where the desired effect is achieved. Frequently, derived chemicals of unknown identity are actually responsible for the result. The anticoagulant rodenticides are only exploitations of chronic toxicity—of the value of multiple exposures no one of which would produce death. In the second class are many of the chlorinated hydrocarbon insecticides in common use and most rodenticides used away from dwellings. The best known of the insecticides are aldrin, DDT, dieldrin, endrin, heptachlor, and toxaphene. Strychnine, thallium, and Compound 1080 are the best known of the stable rodenticides. Residual insecticides have a particular value in the control of disease vectors and soil insects. The word "residual" is not normally applied to rodenticides, but their stability equals or exceeds that of the so-called "residual" insecticides. The most serious hazards to man or animals come from this second category of exceptions.

The first critical aspect of residue hazard is the survival time of stable chemicals in soils or on surfaces. This is normally described by a percentage of loss or "disappearance" over a period. A convenient expression is the residual "half-life," which describes the time required for half the residue to disappear under normal conditions.

TABLE 16

Residue half-life values of a few commonly used, persistent insecticides (from Gunther and Jeppson, 1960)

Chemical	Crop type	Half-life (days)
DDT	Citrus	50
	Peaches	11–15
	Alfalfa	7
Dieldrin	Citrus	60
	Apples	6
	Peaches	6
Parathion	Citrus	78
	Apples	3–6
	Alfalfa	2

A few half-life values (in days) for persistent insecticides on foliage or fruit are shown in Table 16. This abbreviated table indicates the wide variability with crop type that occurs in the survival time of specific insecticides. A list of this kind is valuable to marketers, who must avoid excessive residues on products for sale. It is, however, grossly misleading. First of all, it says nothing about where the residues go or what other chemical compound they may form. Secondly, it assumes that the rate of disappearance is equal throughout residual life—which is by no means true. Thirdly, it is based on single values during single growing seasons. Fourthly, and most important, it gives the impression that these common chemicals are relatively short-lived. In other situations their residual lives can be much longer. There is a common tendency among agricultural specialists to consider a specific residue problem resolved when marketing standards can be met. Biologically, however, this approach resolves rather little.

More important is the residual reservoir provided in the soil. A large part of this reservoir has been derived from the overlying foliage and has been added to the amounts already in and on soil. It is the life of these residues with which we are most concerned. In general, residual chemicals persist much longer under the foliage than on it. Although fractions do indeed "disappear," repeated applications, more frequently than not, restore these missing fractions and add more to them. A build-up of amounts of the residues of stable chemicals in soil has become the rule in crop soils. Moreover, the soil ultimately provides the source from which transformation of chemicals takes place. Not only do amounts of residues increase, but their structure changes. These alterations are not always in the direction of chemicals of lesser toxicity. The limitation of marketing standards in assessing residue hazards is not better illustrated than by these changes in amount and nature. Chemical transformation is not confined to the soil; biological processes arrange for further transformation.

We can divide the question of survival of residual chemicals in soils into two categories: rate of breakdown or disappearance from soils, and build-up or accumulation of chemicals in soil. We can then logically proceed to the next questions: What happens to chemicals in soils, and how can they be transferred from their sites of deposition?

The Survival of Residues in Soil

The disappearance rate of chemicals should be calculated following single exposures. Otherwise repeated treatments obscure the actual rate. Since most treatments are repeated, it is sometimes necessary to estimate rather crudely what the total amounts applied have been and then make annual assessments of residual amounts, from which an approximate rate of disappearance can be made.

DDT, BHC, chlordane, dieldrin, and heptachlor last for long periods in soil. Plots treated experimentally with the unusually high amount of

100 pounds of DDT per acre in 1947 had a residue of 28.2 pounds per acre in 1951. If we assume annual decrements to be equal, this is an average yearly loss of about 18 per cent of the total or about 27 per cent of the residue remaining for each year (Allen *et al.*, 1954). After 3 years BHC applied at 100 pounds per acre was reduced by about half, an annual rate of disappearance of about 20 per cent. Fleming and Maines (1953) reported a loss of about 10 per cent per year over 8 years in areas heavily treated with DDT for Japanese beetle. Lichtenstein (1957), in a study involving 14 orchards, reported a recovery of 26.6 per cent of the total amounts of DDT applied as sprays for a 10-year period. MacPhee *et al.* (1960), in a five-year study, placed the respective average annual losses of DDT and BHC at about 12 and 16 per cent. Lindane and aldrin residues in loam soils did not survive as well; respectively, after two and one-half years, only 10 and 3 per cent were detectable. The possibility of alteration of aldrin to dieldrin could not be checked for no dieldrin analyses were made (Lichtenstein and Schultz, 1959*a*). No heptachlor was recovered from soils treated 9 years previously, but the conversion product—heptachlor epoxide—was present at about 5 per cent of the original application rate. Under the same conditions, 15 per cent of a single chlordane application was detectable after 12 years (Lichtenstein and Polivka, 1959). In the fire ant control-program area, 12 per cent of a single application of granular heptachlor was recoverable in soil after 6 months (Barthel *et al.*, 1960). Heptachlor is expected to be insecticidally active in this program for three to five years.

Other studies could be cited, but those above establish that several common chlorinated hydrocarbon insecticides do in fact exist for long periods in soil. Judgments based on residues on marketed foodstuffs give an erroneous impression of the longevity of these chemicals. Even the organic phosphate compounds normally thought of as "non-residual" will survive in small amounts for several years (H. P. Nicholson, 1962).

The fact that residues do survive is compounded by addition of chemicals in repeated applications, leading to accumulations of residues in soil. Accumulation is particularly important because it can be a hazard to plants growing in treated soil and because it renders more likely the transfer of toxic chemicals out of soil by animal carriers or by physical action.

The increase in soils of arsenic residues resulting from heavy repeated doses of arsenical insecticides has been observed for a half-century (Boswell, 1952). Accumulations of residues reached astounding levels in some crop soils. Levels were particularly high in soils beneath orchard trees and in soils dedicated to cotton culture. Almost all residues were confined to the top few inches, and established plants whose roots penetrated well below the cultivated layer showed little or no effects of these excessive amounts. However, vegetable crops fared poorly in soils heavily contaminated with arsenic, as did cover crops in orchards. Efforts to replace old orchard trees with young usually failed. Attempts to re-use orchard lands to produce cereal, forage, or vegetable crops proved economically disastrous over wide areas. The problem pyramided with increasing resistance of insects to arsenicals, and, particularly in apple orchards, this resulted in heavier, more frequent arsenic applications. This pattern was particularly marked in the Pacific Northwest, where some areas had accumulated amounts up to 1400 pounds of arsenic trioxide per acre. Legume crops became progressively poorer; alfalfa and beans often died on high-arsenic tracts although they thrived on immediately adjacent sites that had no spray residues. Several years of natural leaching and chemical decomposition were required before such common crops as rye, potatoes, tomatoes, beans, and peas would grow acceptably well. The steadily increasing rates of application resulted in amounts on marketed fruit that finally became inacceptable. Great Britain would not buy American apples. The hazard in arsenic accumulations ran the full gamut from soil to marketed product, and led, deplorably slowly, to major changes in the production, processing, and marketing of fruits and vegetables. Economic losses were staggering to growers farming land on which accumulations were excessive. The effects were widespread and occurred over many crop types, and only after many years have the residual effects been ameliorated.

Arsenical insecticides are now used much less commonly. The experience with them over three decades, however, set the pattern for current residue studies and legislation. The appearance of DDT in 1945, followed in rapid succession by several other stable insecticides, lessened the problem with arsenicals but initiated a new set. Insecticides have never been employed so widely as in the last decade. Those used in greatest volume are also stable and accumulate in soils. The best known, of course, is DDT, and it provides an excellent example of the propensity to accumulate in soils.

The possibility of DDT accumulation in soils from spraying is more likely in orchards and with crops where the green plants are turned under and incorporated into the soil after each harvest, as is general with potatoes and sweet corn (Ginsburg and Reed, 1954). In apple orchards, for example, soils may accumulate 30 to 40 pounds of DDT per acre annually. In one study of 12 apple orchards, DDT, after six years of use, ranged in pounds per acre from 35 to 113 under trees and from 26 to 61 between trees. In corn soils the average DDT found was 11 pounds per acre, ranging from 5 to 19. Potato soils averaged 7 pounds per acre. A cranberry bog contained 34.5 pounds of DDT per acre. Values of this kind have been recorded in many situations. It is reasonable to conclude that accumulations will be the rule when crops are treated repeatedly with stable insecticides.

Accumulations do not seem likely in most forest-insect control operations or in any crop situation where application is not unduly great or frequently repeated. Thus, residual toxicity following single or infrequent applications derives from the inherent stability of the chemical or its breakdown products. One might wonder, however, about several special cases in which treatments are relatively infrequent but in which application rates are high. Current examples are provided in the campaigns to control Japanese beetle, imported fire ant, white-fringed beetle, and Dutch elm disease. These control programs call for high doses of chlorinated hydrocarbons (dieldrin, heptachlor, aldrin, or DDT) intended to survive for long periods and not uncommonly requiring retreatment at annual intervals.

The Alteration of Residues in Soil

Although residues in soil do not normally penetrate far belowground, insecticide application to the surface is normally followed by movement of some residues belowground through rainfall and other natural forces. Downward movement is aided by cultivation, and in some practices it is the rule to plow or disk under a surface application. The preponderance of residue accumulations appears to be confined chiefly to the top one foot of soil, and within this depth most residue is found within the cultivation layer (4 to 6 inches). In the fire ant control program most heptachlor residues have been found within the top inch of soil (Barthel *et al.*, 1960). It is clear that humus layers fall within the zone in which most residues occur. Adverse biological reactions that follow residual soil contamination can usually be traced to these few inches of upper soil layers. It is from here that both plants and animals acquire tissue residues, and from here that residues are carried off by runoff or irrigation waters. In other terms, while residues move into soil from the surface, they do not penetrate deeply enough to obviate biological hazard.

It has been generally assumed that disappearance of residues from soil results from chemical breakdown into nontoxic components, from erosion and dissipation, or from binding with soil elements. Actually, little attention has been given to the ultimate disposition of soil residues. Most concern has been directed toward the surviving toxic fraction of the parent chemical. A new phenomenon has shaken dependence on this simple measure. Breakdown or conversion products have been found to be more toxic than, or to complement the toxicity of, the parent residue. The best known cases are the two chlorinated hydrocarbons —aldrin and heptachlor—so widely used in grasshopper, Japanese beetle, and fire ant control campaigns. Millions of acres have been treated with these compounds, which in nature are converted to more toxic derivatives.

Several years ago the conversion of aldrin in soil to the more toxic dieldrin was first noted. This chemical change has since been repeatedly confirmed (Gannon and Bigger, 1958; Lichtenstein and Schultz, 1959*a*) and found to occur widely. Aldrin itself disappears rather rapidly from soil, but its conversion products remain. One study showed that, after several years, six to twelve times more dieldrin than aldrin could be recovered from aldrin-treated soil. Dieldrin formation peaked at about two months, and its residues degraded at a slower rate than aldrin.

A similar picture can be drawn for heptachlor. Conversion to its more toxic epoxide has been confirmed repeatedly. This disturbing consequence led to the establishment of a zero tolerance for heptachlor residues in foodstuffs. Heptachlor is not an uncommon insecticide. The widest single use of heptachlor was in the fire ant control program, including several million acres. At the initially recommended rate (2 pounds per acre) applied to the soil surface, 12 per cent of the insecticide remained in the top inch of soil six months after treatment. This percentage was made up of two and one-half times more of the conversion product—heptachlor epoxide—than of heptachlor (Barthel *et al.*, 1960). The epoxide, incidentally, is far more toxic to the fire ant than is the parent chemical.

Wide-scale use of both aldrin and heptachlor had been made before this conversion phenomenon was discovered. Application rates, hazard precautions, and residue tolerances had been based on the primary chemical. None of the original recommendations should logically be continued in force.

The Transfer of Residues from the Treated Area

In many places throughout the text, reference is made to mortality or lesser effects, of which contaminated soil and plants can be traced as the primary source. Broadly speaking, the vehicles for transport and secondary contamination may be classed as physical and biological.

Toxic residues are carried primarily by water from a treated area to an untreated area. The chemicals need not be dissolved. Contaminated surface-irrigation water and rain runoff have been shown to transport stable chemicals from land into streams and ponds (Nicholson *et al.*, 1962). The usual indication that such transport is taking place is fish deaths. DDT, toxaphene, heptachlor, aldrin, dieldrin, and endrin have all figured in this kind of loss (fifteen streams in northern Alabama polluted by cotton insecticides, entire watersheds contaminated by DDT in forest-insect control, heptachlor leached off into farm ponds in the South, endrin carried into ponds by rain, and so on). Less noticeable effects must occur, but are seldom studied. The U.S. Public Health Service has begun direct measurement of chemicals in major streams in the United States (Tarzwell, 1959). DDT to levels of 0.02 part per million was found in the Mississippi at New Orleans. Trout from Rocky Mountain streams far from sprayed areas had DDT in their tissues (Cope, 1961). Marine fish frequently show residues (Kennedy, 1963). No one can say precisely where these residues came from. Pesticide residues have turned up in other unexpected places. Collection of several species of waterfowl from north of Great Slave Lake in Canada—over 500 miles from the nearest known insecticide applications—revealed residues in almost half the adult birds analyzed. All duck eggs tested (four clutches from incubating scaup) contained residues, averaging 2.2 p.p.m. of DDT and metabolites (George, 1963). It is probable that the amounts of residues will increase, and it is probable that we will continue to be surprised when we find them. The Miller Bill allows a certain amount of residue to leave the land on foodstuffs. It cannot well regulate other routes by which residues leave the land, except by reducing the amounts which are applied to the land.

More important than the physical is the biological path taken when residues leave the area of their deposition (Entomol. Research Div.,

1959; U.S. Fish and Wildl. Serv., 1962). The possible pathways are so numerous and varied as to make classification difficult. They are, moreover, complicated by differing values and by the jargon of specialized fields. Let me outline the main routes of residue transfer, emphasizing that the categories are not mutually exclusive. Many of the categories are sufficiently important to justify detailed treatment in other chapters; these chapters are identified by references within the corresponding section.

SINGLE-LINK RELATIONSHIPS IN BIOLOGICAL TRANSFER

Incorporation of residues into plant tissues.—Some residues are intended to be circulated throughout the plant, as follows the use of some herbicides and the systemic insecticides. Other insecticides may become distributed generally throughout the plant but are not applied for this purpose. Most either adhere adventitiously to the outer surface or, if absorbed, are incorporated into particular sites near the surface. The root hairs and the waxy cuticles of fruits such as apples and citrus are favored sites of concentration. Only adventitious adhesion to the surface can be removed by cleaning.

Direct animal contact with contaminated soil and plants.—Animals may contact toxic residues on plant or ground surfaces and in soil. Physical contact with contaminated surfaces results in chemical assimilation primarily through dermal and respiratory routes of entry (for example, mortality among ground-dwelling birds in the fire ant control area).

Ingestion of plant foods containing residues can result in poisoning symptoms (the usually observed result, such as wood pigeons or ducks dying from ingestion of treated seeds) or low-level storage in tissues. Many animals carry residues in tissues below symptomatic levels (arsenic in orchard birds; DDT and aldrin in waterfowl).

Symptomatic expression of low-level contamination may be delayed until adverse environmental stresses give them significance. The toxic threshold depends not only on amounts of residue but also on physiological state. An example is delayed mortality in trout (without increase in tissue residues).

MULTIPLE-LINK (FOOD-CHAIN) RELATIONSHIPS
IN BIOLOGICAL TRANSFER

Biological carriers of residues.—This type of transfer involves two or more carriers and may be expressed in several different situations.

(1) Secondary poisoning (see Chapter 19): Ingestion by animals of poisoned animals results in secondary contamination, possibly leading to death, to effects on their young, or to storage of residues in tissue. Secondary poisoning is common with rodenticide use (coyotes and carnivorous birds eating poisoned rodents) and probably occurs more commonly than noted among fishes and birds that consume poisoned insects. The poisoning of young birds through storage in eggs of residues derived from parental residues is well documented.

(2) Long-distance carriers: Chemical analyses of migratory birds have revealed just how far residues can be taken from the land where deposited. Waterfowl frequently contain residues of one or more chlorinated hydrocarbon insecticides in their tissues. Sampling to date has shown this kind of contamination to be more common among waterfowl at localities where mortality was also observed (U.S. Fish and Wildl. Serv., 1962). Woodcock in New Brunswick forests during the breeding season have been shown to contain heptachlor and its epoxide in tissue (average 0.18 p.p.m.), which could only have been obtained on their wintering grounds in Louisiana (within the fire ant control area). It is probable in most instances that the source of residue was contaminated food animals (biological carriers).

(3) The human case—residues on marketed foodstuffs: Most animals have narrower food predilections than human beings. Residues may be present on a wide variety of plant and animal foodstuffs. The relationship may be simple, as in the case of direct consumption of contaminated plant foods. Residues may be adventitious, the remnant fraction not removed by washing or processing; or they may be incorporated into the waxy or oily parts of plants, in which case washing does not remove the residue. In most cases any residues are in trace amounts. There is a likelihood of greater residues in animal products. The domestic-animal link interposed between contaminated plant and human ingestion allows storage and accumulation of residues. Residues may be present in the meat, milk, or eggs of the carrier. Residue levels in meat or milk often exceed legal tolerances. In one survey, for example, 60 per cent of some 800 milk samples randomly collected throughout the country were found to contain residues of chlorinated hydrocarbon insecticides; by law, none can be present (Clifford, 1957).

The human being is normally the end of a food chain; fortunately rather little use is made of human tissue. As noted in Chapter 12, however, residues do accumulate in human tissue and in human secretory products. The processes differ little from those in domestic mammals.

The significance of these residues to the human carrier or to nursing children, if transmitted, is conjectural, but their presence is well established and seemingly unavoidable.

Biological concentrators of residues (see Chapter 20).—I have alluded to the transferral of residues from soil to plants, to consuming herbivores, to carnivores, and, finally, to the secretory products of mammals. Stable chemicals may be carried along this food chain, with or without increase in residues. Living organisms may, however, both store and increase ingested residues in their tissues, and yet, at accumulated levels, not seemingly suffer harm to themselves. Algae, earthworms, and fishes are examples. We may term these "nonresponsive concentrators." If a food chain contains a direct series of nonresponsive concentrators, a point will inevitably be reached at which concentrations will mount to levels inducing response. Thus far only warm-blooded animals have been demonstrably killed through this chain of events. The "warm-bloods" also store and increase amounts of residues in tissue, but, unlike earlier concentrators in the series, they do respond to the increasing concentrations of toxic residues. We may call these animals "responsive concentrators." Wholesale mortality has resulted from this sequence. The concentration potential inherent in the transfer of chemicals through a succession of trophic levels can be appreciated only by the ecologically oriented person who is fully aware of the complex animal relationships leading to the succession of concentration levels. (For amplification, see the accounts in Chapter 20 describing this phenomenon in the Dutch elm disease program, in the control of the Clear Lake gnat, and in the fire ant control program.)

It has been my purpose throughout this chapter to show that legislative and advisory preoccupation with pesticide residues in marketed foodstuffs grossly distorts the entire residue problem. No one of us wants less protection against contamination of foodstuffs, and a good many of us want more. But residue legislation is leveled only at marketed foods, and specialists and advisers maintain the curious illusion that current recommendations actually control residues. Limitation of concern to the edible, processed fractions of plant and animal tissue ignores these facts: the soil receives more contamination than do plants growing on it; a biological product can be produced only in a living community; crop plants and their residues are transitory, returning to the soil after a growing season; residues are by no means confined to their sites of deposition; animals are capable of transporting residues away from a treated area (sometimes thousands of miles); residues can be stored and concentrated in living tissue, sometimes yielding profound effects. In brief, residual contamination is general, and exposure is a good deal more complicated and extensive than is usually conceded.

14 · CHANGES IN THE BIOLOGICAL LANDSCAPE

The occupation of land by civilized man has always led to profound changes in biotic composition. The simplest of these changes is the reduction in numbers of native species directly inimical to his interests. The large predatory mammals have been totally eliminated by man wherever the terrain has been suitable for his habitation. Extinction or near-extinction has also been the result among many native species exploited as food or sport. The heath hen, bison, and passenger pigeon are examples. A few native species have adapted remarkably well to the new environments created by man, increased in numbers, and widened their distributions. Deer, coyotes, opossums, blackbirds, and Colorado potato beetles are examples. In all these cases the likelihood of spatial or numerical change depends on the biological adaptability of the particular species. Some are favored; most are not.

On the other hand, introduced species normally find themselves with a particular advantage not shared by their native counterparts. In a biological sense this derives from a lack of the competitors that usually reduce their numbers. The term "balance of nature" describes a dynamic equilibrium of the biota in which both species content and numbers fluctuate little and in which any major change is gradual. Human exploitation reduces the number of species, creates environments favoring the multiplication of a few, and leads to rapid changes in the biota. Under these conditions, introduced plants and animals freed of their normal checks on increase become even more abundant and more widespread in distribution.

The Colorado potato beetle, for example, normally associated with wild species of *Solanum* in the Rocky Mountains, shifted its preference to cultivated potatoes, became more abundant, and made its way to the eastern seaboard, following the route provided by intensified cultivation. On its eastward trek it became sufficiently common to qualify as a pest. Nonetheless, its depredations in this country never equaled those that followed its establishment in Europe. There it continues to be one of the most serious pests of the European continent. Only a slight change in the cast of characters and in geographic patterns gives us such destructive insects as the Japanese beetle, European corn borer, and gypsy moth. Plants also share this pattern. Some 15 million acres in the Rio Grande Valley, once lush grassland, are now occupied by mesquite (Carter, 1958). Within recent memory, excellent grazing land has been reduced to poor, largely through the abuse of overgrazing. The pattern is by no means limited to south Texas; millions of acres in other sections of Texas and in neighboring New Mexico and Arizona have followed the "primrose" path to lesser utility. In a lifetime the rangeland of California, once dominated by native perennials, has become composed largely of annual grasses of Mediterranean origins. A large part of range "management" is an attempt to restore perennial species. Cactus in Australia or South Africa, lantana in Hawaii or Rhodesia, Klamath weed in California or St. John's wort in Australia, and gorse in New Zealand typify the introduced plant species that have spread, become established, and increased their numbers sufficiently to be labeled "pests."

Man has not depended on native crops for his sustenance. His food plants are also introduced. Being greatly changed from their ancestral character by selection, they are no longer subject to the natural processes that render them able to compete effectively with other plants and animals. The practices of planting large areas with single species of plants, in either agriculture or forestry, and of encouraging greater plant growth through cultivation, irrigation, and fertilization create especially favorable environments for species that also find these conditions favorable. Man is responsible for the conditions in which pests flourish. By far the greater part of pest control is directed toward correction of imbalances of man's own creation.

Imbalances may be the logical derivatives of necessary practices, or they may come from ignorance. Whatever their source, they are intensified by modern agricultural practices and the ease of present transport. It is not my purpose here to dwell on the effects of man on environments. An interested reader should refer to the excellent summary volumes of Elton (1958), Darling (1955), E. H. Graham (1944), and Thomas (1956). In the present context I must emphasize a terminal point of view. Perhaps a series of topical questions, some of which have already been discussed in other connections, will illustrate the position of pest control in relation to biotic change. Does pest control accentuate the existent trend toward simplified biological environments? Does it aid in the dispersal of pest organisms? Are we presently reducing desirable species of plants and animals to extinction through pest control? Are we unknowingly producing changes in biotic make-up over large areas? Or, in still more topical terms, are we producing "biological deserts"? Is plant control on roadsides and utilities rights-of-way a hazard to the native biota?

One might pose other questions. The answer to all those above is a qualified "yes." But answering questions of this character in precise terms is impossible.

The first limitation is one of values. The majority of people neither pose the questions nor worry about longer-range consequences. Among the minority who do, there exists a conflicting set of values. On the one hand, the extraordinary thoroughness with which man (particularly European man) has pursued his exploitation of environment has resulted in a material standard unparalleled in man's existence. The pattern continues to spread throughout the world and is looked on by most peoples as desirable and progressive. On the other hand, the ecological domination by man should be attended by moral responsibility toward nature. An enriched "way of life" includes more than the material standard we are capable of achieving. Moral responsibility does not solely rest on preservation of objects of beauty, curiosity, or sport. The "ecological conscience" (after Aldo Leopold) also includes the responsibility to keep in being the productive sources on which continued material well-being depends. In the minds of many of us there is doubt that the thoroughness of man is matched by his wisdom. Conservationists can call to mind hundreds of examples that raise this thought from the realm of doubt.

The second important limitation is technical. We have few good measures to establish the significance of ecological change. Those that we have are too rarely used in practice. Change is clearly seen, but we become alarmed only when declines in productivity or in favored species become irreversible (or nearly so). We do not have the knowledge (or the wisdom) to predict the long-range consequences of our exploitation. We can depend on historical examples of man-induced extinction for object lessons; these are useful in substantiating the conviction that for continued welfare man and environment must live in harmony. But history offers no examples of the rapid changes that modern technology has enabled.

Nor can we be depended upon to use knowledge wisely! Kuenen (1961) noted that "... a rational analysis soon teaches us that irrational factors govern many of our actions even today." The knowledge on which we must depend for wise programing in ecological management should come from scientists who are aware of the complexity of the problem. Most often, studies—even good ones—are imperfect in that small fractions of a problem are studied, and rarely is an attempt made to link these fractions with the entire problem. Too many scientists today thus limit themselves. The reasons are many; a few are offered here.

Many scientists are single-minded, not concerned with large concepts or integration of facts from widely diverse disciplines. Academic training is in part responsible, as is the degree of specialization within sciences. The manner of organization of many biological "service" divisions of government contributes much to this limitation. Research in these agencies is usually short-term, limited to immediate problems, confined to the fraction judged to be economically significant at the moment, strongly conditioned by special-interest pressures, and in any event inadequately funded. It is small wonder that scientists in agencies cannot generally be looked to as sources of the information and full-spectrum appraisals required in ecological management.

There is, perhaps, a more serious charge that can be made against scientists who are able and free to make the studies we desperately need. Particularly in the United States, few eminent biologists are attracted to the study of ecological problems of social significance. "Disturbed" areas are not sought for study. Seemingly overlooked is the fact that little now remains undisturbed by man. Grave social decisions rest upon our comprehension of man-caused ecological disturbances. The best minds should be turned to concentrate on it. That some change in this attitude has recently taken place is attested by the volumes I previously cited. The authors of those works are pioneers in an intellectual land requiring full settlement.

Further change is attested by the recent activities of both state and federal agencies. At both levels of government, appraisal committees have been formed to examine the ecological and other hazards of pesticides. The Governor's Committee on Pesticide Residues in California is one example. Another at the federal level is the President's Scientific Advisory Committee report on pesticides. The Congress too and the National Research Council have made appraisals limited to certain phases of man-induced changes caused by pesticides. Even so, only beginnings have been made. The extent to which modern man modifies the environment he occupies has not yet been fully appreciated.

These explanatory comments may seem to beg the questions I have posed. They are offered to orient a reader unfamiliar with the general state of studies in field biology. They also amount to an apology that precise answers of general application are not possible.

Does pest control accentuate the trend toward simplified biological environments? Yes, but in most biological categories it has yet to become the primary cause. Only among predatory mammals has pest control been a major reason for reduction in numbers or extinction. Among pest insects no native species to my knowledge has been reduced to extinction. Very few insects—all introduced—have been eliminated from large areas. Examples are *Anopheles gambiae* in Brazil, Mediterranean fruit fly and screw-worm in Florida, and date palm scale and Khapra beetle in California. Reduction of numbers is frequent, sometimes nearly to the point of extinction, but the cause is normally found in aspects of man's presence other than deliberate control. Biological control, by its nature, reduces the numbers of individuals but enriches species representation. Potentially, chemical control is capable of producing effects on the biota as profound as the habitat changes brought to bear by the presence of man. In limited crop situations it already produces great changes. I shall illustrate in detail the character of these changes in animal numbers and species in a later section (see

Chapter 21). Herbicides abet the maintenance of monocultures in agriculture and forestry. Westhoff and Zonderwijk (1961) consider, for example, that about one-third of the plant species in the Netherlands are threatened by rural use of herbicides. The drive for "technical efficiency" accounts for the success of herbicides in eliminating competing species from cereal and forage crops. Selective herbicides can effectively create single-species stands in forest plantations and on rangelands. There is no evidence that soil treatments with pesticides accomplish anything but temporary alleviation of high numbers; cultivation, of course, can eliminate an animal species locally. In summary, pest control does contribute to biotic simplification through reduction in numbers and shifts in species domination. These are its purposes. But, with the exception of large mammals in populated areas, no native species of plant or animal seems to have been eliminated by pest control practices *per se.*

Does pest control aid in the dispersal of pest organisms? Yes, but in a spatially limited way. Organisms that survive and adapt to pest control procedures become reservoirs of dispersal. Trappers believe that aggressive control campaigns have selected for wilier coyotes that are extending their range. Bait shyness in rodents and pesticide resistance in arthropods result in adapted populations able to re-invade or to establish populations in new areas commensurate with their innate ability to disperse. The greatest potential reservoir of dispersal is in arthropod species, once uncommon, that became abundant following the use of insecticides (because of removal of competing insects by nonselective chemicals). There are to my knowledge no quantitative studies of the extent to which these new populations have dispersed. There is, of course, abundant evidence that changes in habitat can lead to widened distributions of animals and plants.

Are we presently reducing desirable species of plants and animals to low numbers or extinction through pest control? I have answered this in part under the first question. Emphasis in this context must be on value—on what "desirable" means. Since most species are considered valuable by someone, the answer is "yes." The coyote, now uncommon in some areas, is often regarded as a valuable aid in rodent control, as well as a picturesque representative of the fauna of the American West (notwithstanding the attitudes of livestock producers and their supporting governmental agencies). As a more pointed example, the fulvous tree duck, it is believed, will be in serious danger if the practice of treating rice seed with aldrin and other toxic chemicals is continued within the bird's limited range in the United States (U.S. Fish and Wildl. Serv., 1962). Numbers of peregrine falcons and other raptorial birds have declined markedly in recent years in parts of Great Britain and northern Europe. Systematic analysis of both flesh and eggs of raptors has commonly shown significant residues of chlorinated hydrocarbon insecticides (Moore and Ratcliffe, 1962). Several common species of birds in Great Britain seem to be declining as well. The presence of insecticide residues in tissues of many of them has been established. But it is not yet possible to affirm that insecticides are a primary mortality factor (Moore, 1962). Repeated reference has been made to losses of insects beneficial in natural control. In the majority of cases, however, the cause of decline in native species must be ascribed primarily to land use and not to pest control.

Does pest control produce biotic changes over large areas? Are we producing "biological deserts"? Yes; pest control, where it assists in simplifying habitats, accomplishes these things. Chemical control reduces the numbers of some desired faunal elements, but very rarely eliminates species. In a few instances there is the strong suggestion that, if continued, present patterns of control, involving repetitive treatments over large areas, will lead to permanent depression of both numbers and species of birds. An example is the loss of insectivorous birds in areas of Dutch elm disease control by DDT. Where repeated treatment is not the rule, permanent depression of numbers seems unlikely. The recovery powers of most species are sufficient to ensure this. Uncommon animals are exceptions. The decline in numbers of the bald eagle is a case in point. Their numbers are rapidly declining, and recent field sampling and feeding experiments very strongly suggest that slowly accumulating pesticide residues have contributed to the drop in numbers (DeWitt and Buckley, 1962; George, 1963). Pest control can also lead indirectly to loss of species. It is conceded, for example, that mastery over the tsetse fly will free large areas of central Africa for human occupation, which, gauged by experience, will lead to the extinction of many unique large mammals now found there. In the not-too-distant future, techniques will be available to permit this mastery. Local successes have already begun the sequence.

Is plant control on roadsides and utilities rights-of-way a hazard to the native biota? Yes, in the sense that species composition is always altered. The advantages in maintenance are clear. Egler (1958) notes that the rights-of-way of the utility corporations alone comprise an acreage greater than all six New England states combined. Unquestionably the increasing demands of transport and communication will lead to greater acreages modified for easy maintenance. Alterations of this magnitude cannot be taken lightly. Rights-of-way are not private lands. Many interests and values are served by them. Their proper management should be a concern of an entire citizenry.

I conclude then that changes in biota are inevitable consequences of intensive land use, that fewer species are present in lands thus altered, and that pest control is normally only a secondary instrument in biotic change. Pest control does produce change, but its usual result is a temporary reduction in numbers. Lest it appear from these conclusions that I do not regard pesticides to be a source of major biotic change, I must offer these qualifications. I have attempted to make general conclusions of the widest applicability. In more restricted ways (though often influencing millions of people), biotic changes caused by pesticides are profoundly important. These changes are, after all, major preoccupations of applied biologists, public-health workers, and conservationists. Hazards to human health and to desirable forms of life, while perhaps not producing immediate biotic changes, should nonetheless be considered part of the general problem. Moreover, my general assessments are based on traditional pest control practices. We have barely begun to see patterns of influence that we can term nontraditional. Mass spraying with nonselective chemicals is only in its infancy; intensive, large-scale, monocultural agriculture and forestry have only begun their probable courses throughout the world; the transferral of insects and diseases from one area of the earth to others must inevitably accelerate, however impressed we might be at the moment with current devices for preventing it; current evidence for the manifold ramifications of pesticide influences in living environments gives only a hint of their future significance; rapidly accelerating numbers in human populations, out of harmony with their food supply, must lead to desperate measures of pest control; more fortunate peoples will increasingly demand higher quality and productivity of foods and fibers, simultaneously insisting on more protection for wildlife, waters, parks, and preserves. Without question, the significance of pesticides as inducers of biotic change will become magnified in the next decades.

V · ECOLOGICAL RELATIONSHIPS AND CHEMICAL CONTROL

Curiously obscured in an era of technological domination is the fact that one living thing depends for its existence on another. Man derives needed energy and pleasures from living things. The goal of crop, livestock, timber, and wildlife managers is the effective channeling of energy from living things into useful products. Pest control at its simplest ensures the well-being of man by protecting his person and by inhibiting those living competitors that would take from him what he wants and needs. It follows, therefore, that both the productive and the protective phases of man's relationship with other living things form a continuum of complex interactions among living things; the scientific study of these relations is ecology.

Empirical manipulation of ecological relations to yield ever-increasing advantage to man has been the established pattern over many centuries of experience. Scientific application of ecological rules with the same goal in mind is new. Already it is yielding a great deal, and promises much more; applied ecology has only begun to make its con-

183

tributions. Some of its contributions may be more easily recognized if the component disciplines are identified. Here are the most important: forest biology, range management, wildlife management, soil science, economic entomology, plant ecology, conservation, human ecology. Each of these disciplines is represented by professional societies, agencies in government, and departments in universities. The fragmentation into specialty disciplines does not alter the fact that all are concerned with the same biological continuum. Linking them all is an inescapable web of ecological unity. The ecological approach to utilization of living resources provides the only sure route to continued safety, productivity, and pleasure. Successful pest control measures always have at base an understanding of the natural relationships of the pest species in question.

The importance of this ecological foundation is not as widely appreciated as it should be. In the present context there are two clear reasons why that is so. One is the recent origin of the formalized science of ecology. Younger still are the management disciplines loosely grouped under "applied ecology." More information and synthesis are required before these disciplines reach full status. Secondly, the impressive achievements of chemical technology in recent decades have clouded perspectives among many applied biologists. Commonly, biologists had come to think that their problems were chemical rather than biological. This inverted perspective led to practices in which chemical agents became not precision tools wisely and sparingly intercalated into living systems, but sledge hammers which indiscriminately affected large elements of these biotic systems. Ideally, to paraphrase A. W. A. Brown (1951), a chemical weapon in biological systems must be used as a stiletto rather than as a scythe.

My purpose here is not to present a general ecological accounting of man's impingement on the living environment. Rather, it is to describe those biological relationships that most cogently show how pests result from crop-production practices and how pest control methods telegraph effects beyond the target species.

Thus limited, ecological relationships in pest control are of three kinds. The first concerns the changes in environment that result in the creation of large numbers of damaging animals and plants. In general, crop and forest land management results in reduction in total numbers of species previously present and in the creation of conditions favorable to excessive multiplication of the few species that can adapt to the new environment. The second relationship involves the need of all animals for suitable foods. The maintenance of uninterrupted food chains free of chemical residues is essential to the welfare of most animal species. The third category deals with population balances, and is more narrowly concerned with the changes in the numerical relationships of species following reduction in numbers of a target species.

15 · THE CREATION OF PESTS

Pests are organisms that, individually, can produce personal harm or that, as populations (the individual having little meaning), deprive us of something we wish to keep for ourselves. Success in pest control is measured by how well we approach full protection of person and property. Too often overlooked is the fact that the word "pest" has no biological meaning. A pest species is subject to the same ecological rules governing the survival of any other form of living thing. What we choose to label "pest" may merit the appellation only because we have ourselves altered its normal distribution, numbers, or relationship to other living things. Thoughtful pest control depends on understanding the extent to which man's living requirements favor certain species of animals and plants. Methods to manipulate the favored habitat adversely are often revealed through this understanding, and, as important, comprehension of natural relationships allows recognition of the point at which nonbiological means of control must be interposed to assure continuing advantage to man.

Three concepts underlie the practical consideration of pest creation and aggravation by man-caused disturbances of the biological environment. These are the community concept in ecology, the simplification of biotic communities, and the derived consequence—the concept of plant and animal weeds.

The Biotic Community

The community concept in ecology is a formal recognition that living things are not randomly associated. Natural assemblages are easily identified by their appearance and distribution. Sophisticated analyses are not necessary to confirm that grasslands, conifer forests, rain forests, woodlands, savannahs, and tundras, with their associated faunas, are natural groups. Their continuance depends on the physical contributions of soil and climate and the biological conditioning that accompanies any occupation of land. Given static conditions, the biological community remains stable. Over time, instability is the rule. In the short range, instability results from cataclysm, examples of which are such diverse forces as hurricanes and forest clearing by man. Over longer periods, entire communities evolve into other types through slowly operating natural processes, examples of which are glaciation and soil formation and erosion. Yet larger aggregations of plants and animals are identifiable on a continental basis. Continental identity of biotic communities is maintained by the effective barriers to natural dispersal provided by the major mountain chains and bodies of water.

The survival of individual communities depends on the integrity of the soil, which is in turn dependent on the plant and animal associations living on and within it. The existence of any species population of animals that is part of the community is contingent on the survival of other species populations of the community, since food and shelter can be provided only by other organic members. There exists, in the numbers and kinds of species in a community, a general balance dictated by competition for the prime requisites for existence—food and shelter. Ordinarily the numbers and kinds of organisms in a stable community fluctuate little. The resulting approximate equilibrium so clearly seen in stable communities gives rise to the popular concept of "balance of nature."

This natural balance cannot be maintained whenever important changes occur in single constituents. In this context it makes little difference what the cause of change might be. A change in soil condition will inevitably result in change in the dependent flora. The resultant modification in plant character (a changed habitat, as animal ecologists are prone to call it) must inevitably lead to different species and numbers composing the fauna. The sequence of change need not start with the soil. If at any point the numbers of a single species are caused to change, the intimacy of the internal relations within a biotic community ensures that other members will change.

Thus far I have noted that a certain order is apparent in the distribution and aggregation of species. The word "community," however widely applied spatially, connotes a predictable form and function among its plant and animal components. The numerical composition of these varied organic components is stabilized within relatively narrow limits by interaction among them, the most important one being competition. It remains, however, to point out that destruction of a community structure is followed by a regular sequence of biotic assemblages leading to restoration of the original community. This ecological *succession* is orderly, directional, and capable of description, and its end products are in many instances predictable. The entire sequence of successional stages is spoken of as the *sere*. Changes taking place within the sere are called seral phenomena. The unstable early stages of a sere are termed pioneer communities, and the relatively stable end product is known as the sere climax or, more often, simply the climax.

From this brief and general summary, one should not conclude that all biotic sequences leading to community formation and stabilization are well understood. It is always easier to describe appearances than to understand mechanisms. These ideas on community characteristics are nonetheless accepted by professional ecologists and should be kept in mind in reading the more detailed descriptions to follow.

The Simplification of Biotic Environments

Reduction in the complexity of the biotic elements in living communities is the natural consequence of converting lands to the cultivation of single or, at best, a few crop species. Attendant upon this conversion is an actual reduction in the number of plant and animal species in the area thus disturbed, the likelihood of greater numbers of the species that remain, and a greater susceptibility of the land to invading species. In successional terms, the effect of land conversion is to push seral stages back toward pioneer, maintaining a permanent subclimax, and hence unstable, community. In terms of agriculture or forestry, the effect is to channel more energy into favored species, to provide greater yields of their useful products. Biotic simplification is essential to the maintenance of crop species everywhere, and in all times has been attended by pest problems. The modern "simplification for efficiency" is perhaps only a special case, but its scope and the manner by which it can now be effected have brought a new set of hazards. Before going into these hazards let us look more closely at the processes by which community simplification can be achieved.

Simplified environments occur in nature. In general, the degree of complexity of the biota is correlated with the rigorousness of the environment. Clear ecological patterns of distribution are discernible. The biota becomes increasingly complex as one progresses from pole to equator, and decreases in complexity as one ascends to the tops of high mountain ranges. We may therefore speak of latitudinal and vertical simplification. In special situations the biota for the latitude in question is yet further simplified by topographic conformations that cause unusual aridity.

Latitudinal differences in the complexity of biotic communities are astounding. In arctic tundra, for example, there may be only a dozen or two permanent species of plants and animals. The short summer accommodates a few more species, particularly of breeding birds. Mosquitoes, midges, and blackflies become abominably abundant during the summer. Migratory patterns are pronounced to compensate for the limited seasonal food supplies. Violent oscillations in numbers characterize tundra communities. The classic lemming cycles, in which five-hundredfold changes in density occur regularly, are paralleled, but with lesser amplitude, by changes in the numbers of avian and mammalian predators. In contrast, a few acres of tropical rain forest may contain more species of trees than are found in all the flora of Europe (Odum, 1959). On the six square miles of Barro Colorado island, a research station in the tropical rain forest of the Panama Canal Zone, there are some 20,000 species of insects, more than are in all of France. Seasonal stability at tropical latitudes obviates the need for migration, and only in special situations, chiefly semiarid areas, does it occur at all. The numbers of individuals in a species are usually much fewer than at middle or high latitudes, and no violent oscillations occur in population density.

Artificial simplification is a convenient term to describe the reduction of species diversification that occurs wherever man uses land intensively. The only human cultures which use the land without simplifying the biota are those of the hunters and the gleaners. It is, however, not biologically correct to consider man an artificial influence in biotic systems. The word "artificial" becomes a more acceptable term if we consider what the biota would be had man not entered the picture.

Since the consequence of man's manipulation of land is always biotic simplification, in self-interest it is pointless to argue that simplification is harmful. We must clear, cultivate, and plant. The important question is how far simplification can be carried without interfering with crop yields or other products of lands. There is, of course, no single answer, and even when the question is applied to particular sites, the precise point of decreasing value will be difficult to identify.

Four processes, all interrelated, are discussed most easily. These are clearing, cultivation, planting of single species, and abuse of land. The last-named is, of course, a special condition, whereas the first three are the normal sequence followed in agriculture.

The simple clearing of forest and herbaceous growth, followed with preparation for cultivation, reduces the numbers of plant and animal species. The means of reduction is obvious enough when one visualizes mixed hardwood forest cut to the bare ground. The sequence of biotic change following clearing is easily apparent, logical, and widespread. But, surprisingly, documentation from careful study is rare. No appraisals of changes in the total biota undergoing alteration seem to have been made at all. Changes in particular groups are recorded more often. In Malaya species of animals are much more numerous in rain forests, for example, than in cleared areas. In Selangor, Malaya, Audy (1956) noted at least 49 species of rodents and insectivores in original forest, but 15 (5 of which were introduced) in deforested areas, and only 5 (all introduced) in built-up areas. Mammalian ectoparasites (trombiculid mites) paralleled this reduction in species—58 in forest, 16 in cleared area, and only 3 in town. Probably 4 of the 19 species occurring outside the forest were introduced. Accompanying the reduction in species number is a usual pattern of confinement and concentration in the disturbed areas.

In another example, the successional return of once-cultivated land to the climax community was carefully studied over several years. We may take the parallel increase in biotic complexity to illustrate the reverse of the clearing process. Abandoned cropland in the piedmont region of Georgia followed five seral stages in its return to climax: forbs, grass, grass-shrub, pine forest, and oak-hickory climax. Breeding passerine birds increased in both species variety and density along this sequence. The change in variety ranged 2, 2, 9, 18, and 20. Estimated pairs per 100 acres increased in this fashion: 15, 40, 136, 158, and 228 (Johnston and Odum, 1956).

Normally, cultivation and planting are so closely linked that separating them is pointless. However, cultivated ground with plants growing sustains a more diversified fauna than does bare ground. In a sense, planting begins a reversal of direction of the path of simplification. In this context, however, the important feature is that the complexity of crop ecosystems never approaches that of the ecosystems they replace. Single varieties of crop species (monocultures) are normally grown on cultivated lands. A monoculture in a diversified landscape, of an area not so large as to exceed the dispersal powers of its animal members, maintains a reasonably diverse and stable fauna. From the standpoint of ecological stability, such monocultures cannot exceed sizes normally characterizing garden or peasant agriculture. American agriculture long ago left this standard behind. The extreme example of monoculture is in cereal crops, which frequently cover many thousands of acres with solid stands. More frequently in other crops, single stands range from 40 to 200 acres. These, too, are of sufficient size to guarantee ecological simplicity and instability. Annual crops, particularly when varieties are rotated, are less prone to catastrophic instability. Longer-lived crops— orchard, forage, and pasture types—when of large extent, are particularly susceptible to the consequences of instability.

Biotic simplification is further accentuated by changes within and around croplands. The maintenance of clean fence rows and field roadways, and of neighboring trees, shrubs, and herbaceous verges, removes reservoirs of complexity where evicted fauna could otherwise easily colonize. The practices of clean cultivation, either by machine or by chemical application, simplify the biota further. An example of the latter is selective removal of broad-leaved weeds in cereal fields to facilitate harvesting.

Artificial simplification is hence the normal and necessary pattern in agriculture. Simplification for efficiency of maintenance and ease of harvesting is a relatively new practice. So also is the practice of large-scale monocultural farming. These elements offer the base for catastrophic instability. Taking together these varied means by which biotic simplicity is currently achieved, we may note that the world-wide trend is toward a crop biota scarcely more complicated than the arctic tundra.

Parallel patterns have on occasion been followed in forestry. The pine stands of the southeastern United States have been maintained by fire and herbicidal treatments as essentially pure stands of large extent. Insect outbreaks have sometimes been linked to the maintenance of such monocultures (Franz, 1961). However, forest ecologists tend to appreciate (perhaps are forced to) the value of diversified plant stands and an attendant stabilized fauna; the long life of timber trees and the uneconomical character of repetitive chemical and mechanical treatments dictate a policy of least disturbance. Various measures, including diversification of stand—recently, mixing conifer stands with hardwood species—are now used to lessen the severity of insect outbreaks.

Range managers, too, have recommended, with increasing frequency, mixtures of plant types for reseeding grazing lands. In this instance, as with many other examples in agriculture and forestry, diversification of plant types may have values other than ecological stability. For example, mixtures of annual and perennial grasses with different ripening times can extend the period that cattle can efficiently graze pasture lands. Whatever the purposes, the value of conscious biotic complication ensues.

Some of the practices I have described may be looked upon as ecological abuses, and indeed are so regarded by many ecologists and conservationists. Nonetheless, there are economic reasons for their being, whether the goal is production or ease of handling. There is a special category of abuse whose consequences are recognized by crop producers themselves. Whenever crop or grazing lands are utilized at intensities beyond their powers to sustain desired plant species, biotic simplification results. The natural grasslands are particularly subject to abuse of this kind. Many early civilizations developed in grassland regions, and faded through their abuse. Even today, and especially in the recent past of the western United States, thousands of acres continue to be converted, at worst, into useless desert or, at least, into land of lowered productivity. Accompanying reductions in plant and animal species have been followed by excessive multiplication of a few, as in the rise of dominance of annual species of grasses on overgrazed rangeland. An increase of rodent numbers is a normal consequence on land of this character and may, in the continued presence of heavy grazing, contribute to further range simplification and destruction. Needless to say, numbers of grazing livestock must be reduced on simplified range-land if wasteland is to be avoided.

The Concept of Plant and Animal Weeds

According to one definition, a weed is a plant in the wrong place. It is a pest because we term it so. Weeds may be well-behaved in their natural habitat, but when their biotic relationships are changed—particularly when the plants are transported to new areas—they develop more vigorously than local occupants. Most species of nuisance plants are not native to the area where they are called "weed." Almost all have been carried to the new area by man, some intentionally. Most of the noxious plants in California, for example, came from the Mediterranean area; a few important species derive from northern Europe. Development of weeds *in situ* is almost always related to man-caused disturbance of land, frequently because of serious abuses of land. Only in rare instances is a native species of wild plant sufficiently hardy to continue serious competition with introduced crop plants if the cropland is well managed.

To the ecologist the patterns in which plant, insect, or rodent occupy disturbed and simplified habitats are much the same. The concepts of biotic simplification and community succession should prepare a non-ecologist for the idea that all are biological weeds, hardy occupants of disturbed environments. In Malaya, for example, Audy (1956) notes that the introduced weed *lalang* (from Africa) occurs in dense concentrations following deforestation. Associated with it are two species of introduced rats, chiefly *Rattus argentiventer*, which in turn is heavily infested by two species of mites that carry scrub typhus. One of these mites (*Trombicula akamushi*) was introduced to and spread over Malaya largely by birds.

The term "animal weeds" was introduced by Taylor, Vorhies, and Lister (1935), who in this instance were describing induced abundance of jack rabbits as a response to overgrazing on rangelands. Studies on range rodents—kangaroo rats, prairie dogs, and squirrels—have repeatedly illustrated their increase following disturbance of the climax, and their decline as climax vegetation is restored.

Whether animal or plant, the following characteristics are normally associated with biological weeds:

(1) They are vigorous in disturbed areas (subclimax communities).

(2) Efficient reproduction and dispersal ensure extended distributions. Ordinarily "weeds" are introduced species.

(3) They are usually transported successfully only from one simplified (man-caused) habitat to another, and their occupancy tends to be confined to the disturbed habitat (they do not invade more complex habitats).

(4) In these areas of occupancy they multiply more rapidly and maintain higher densities than they did in their native habitat, or than do their counterparts in adjoining complex communities.

(5) The abuse of land accentuates their multiplication which frequently leads to catastrophic devastation, often with far-reaching adverse effects on the capacity of the land to be regenerated.

(6) Removal of biological controls, as a consequence of simplification, renders excessive multiplication and population instability more likely.

(7) If all favorable conditions are met, high densities of the noxious species are predictable. Equally predictable is high resistance to control campaigns by man. The compensatory mechanisms are many. In plants, seed dispersal, dormancy, or vegetative reproduction may ensure survival and quick recovery. Among pest rodents, the most common animal weeds, any sharp decline in numbers is followed by greater reproductive success. Two factors in this success are increased litter size and more frequent estrous cycles.

Awareness that biotic simplification increases both crop yields and pests is not new. The realization that effective ecological management may result in adequate crop yields and lowered pest densities is relatively recent and not widely accepted. Early quarantine practices were empirical recognition that simplified environments were particularly vulnerable to introduced pests. Manipulation of living counter-pests, now labeled "biological control," is centuries old, but only in this century has it gained widespread acceptance. Integrated control as a special aspect of ecological management is a relatively new concept, which in ecological terms means maintaining as much biotic complexity as is consistent with high yields of the desired product, and using supplemental chemicals where needed. In several production-management disciplines—soil conservation, range and wildlife management, and forestry, as examples—the practice of maintaining biotic diversity, and of inducing it, has come to be a basic feature. Only in the last two or three decades have these disciplines gained stature.

The ecological unity closely binding the sciences applied in agricultural production is not sufficiently appreciated by the practitioners themselves. Fortunately, a favorable direction has been established in the last decade among professional ecologists, and the movement is rapidly picking up force. The time may be very near when total ecological management will have gained as much acceptance as have the varied, fractionated management disciplines that are concerned with it. Resistance to full acceptance of ecological management among people in production, distribution, and regulation arises chiefly from failure to recognize the biological basis or the practical advantages of biological manipulation.

Analyses of the biological basis for land uses are summarized by Elton (1958) and E. H. Graham (1944). Let me point out again that ecological management depends on reversing the trend toward environmental simplification. Its practical measures rest upon retention of diversity and the conscious restoration of more complicated ecosystems. Precise methods have been repeatedly described in other contexts throughout this book. For statements of economic advantages accruing from the application of ecological knowledge, a doubting reader may refer to both Elton's work cited above and a later summary of E. H. Graham (1947). The consequences of this application serve many values, by no means the least of which is better pest control.

The word predator comes from the Latin *praeda*—prey. Biologically as well as etymologically the words have a common root, and one is the logical derivative of the other: There can be no prey without a predator. No living community lacks predation, and only in environments dominated by man is the biological importance of predation diluted. Predation is therefore biologically normal, ubiquitous, and with relatively few exceptions, essential to man's welfare, although that is not immediately obvious to the non-biologist. Yet, it is largely from these exceptions that there has come a grossly distorted picture of what predation is, how it functions, what its beneficial and detrimental actions are, and how the competition with man's interest can best be met. My task here is to describe these varied phases in ways which may lead to understanding the full meaning of predation as it relates to pest control. Unfortunately I cannot do that completely successfully— nor can anyone—for ignorance, special interest, and conditioned values ensure disagreement. Few subjects in biology lead so readily to controversy as do predator-prey relationships.

Biologically, a predator is a flesh-eating animal. Occasionally the concept of predation is extended to plant-feeding animals as well, but this use dulls effective definition: It is merely another way of saying that animals depend on plants for food, whether or not intermediate links intrude to convert plant tissue to flesh. The value of such an approach is that it facilitates description of the path of energy dissipation in ecosystems and the efficiency of energy conversion at points of transfer along this path. As important in ecology as is this approach, our emphasis must be on individual species and the size of their populations, for these are the present measures of success and hazard in pest control. When, therefore, I speak of predation, only a relationship between animals is intended. I should add, however, that the parasites of animals can normally be included in this definition. They too depend on animal tissues; only their means of dispersal, of "capturing" animal food, and their effects on whole organisms differ. Moreover, parasitism and predation, viewed from the standpoint of entire populations, often produce results of the same kind.

Two considerations underlie all practical discussions of predator-prey interactions: prey specificity and control of prey populations. Expressed as questions, we might ask: Is a given predator limited to a certain kind of prey? Does the predator control the size of its prey population? Neither question can be generally answered with "yes" or "no." However, I hope that acceptable qualified answers will emerge from the discussion to follow.

The Biology of Predation

The components of predation are many and varied. Nonetheless some general categories are recognizable. First among these pertains to the feeding specificity of the predator. Facultative feeders are those able to live under different conditions, whose type of prey is optional, and whose feeding habits are opportunistic. Obligate feeders, in contrast, are bound closely to particular environmental niches and food types. Among carnivorous mammals, for example, bears are essentially facultative feeders. Their diet includes roots, berries, insects, fish, rodents, carrion, and livestock. The choice of foods is dictated somewhat by preference, but more often by the opportunities afforded by season and locality. Coyotes depend basically on rodents and rabbits at any season of year, but their diet varies greatly with opportunity. They may be termed partial obligates. Mountain lions, in contrast, depend primarily on deer for food, and although other items such as rabbits will be eaten, the welfare of mountain lions is tied to the numbers of deer. For practical purposes, they are obligate feeders. The same kind of comparisons could be made throughout the vertebrates. Insects tend to be more narrowly adapted, and in fact the best examples of obligate feeding among animals occur among insects. For illustration, I need only to point out that the success of biological control operations depends on the degree to which a predatory or parasitic insect is restricted to a single prey species. A completely obligatory predator is preferred; the welfare of one species thereby depends solely on the other.

Many factors other than feeding predilection affect predation (Holling, 1961). The basic variables are the densities of predator and prey —inevitable features of any predator-prey situation. Other factors derive from these.

The number of prey destroyed by predators is, of course, the product of the number killed per predator and the number of predators present. This twofold nature of predation was recognized by Solomon (1949), who proposed the terms "functional response," to apply to prey consumption, and "numerical response," to concern density of predators.

Published information on the functional response of predators to prey density deals largely with insect parasites (a special type of predation). In all instances parasites attack more hosts as host density increases. But this direct relationship of attack to host numbers does not continue indefinitely. At higher densities, the relative intensity of attack decreases, so that the curve describing the relationship tends to flatten out. The relationship is curvilinear, appearing on a graph as a simple arc flattened at its top end, when density is charted on the abscissa and percentage of parasitization on the ordinate. Functional-response curves of this kind are not limited to parasites. The same kind of thing has been shown among predatory insects.

The functional response of vertebrate predators to density of insect prey seems to be of a different sort. Small mammals (*Peromyscus, Sorex, Blarina*), for example, are important predators of cocoons of the European pine sawfly. Both experimentally and in nature, individuals of each species attacked more prey as prey density increased. Unlike the response curve of insect parasites, the functional-response curve of these insect predators was "S-shaped," or sigmoid. It appears therefore that vertebrate predators attack scarce prey by chance, but in some way develop the ability or inclination to find a greater proportion when the prey become more abundant (Leopold, 1933). The number of predator attacks is thereby associated with greater skill acquired through learning. Here, too, prey taken cannot be indefinitely increased, simply because there are limits to the numbers of predators and to the food they require. The response curve at this upper limit must thereupon flatten out to form an upper plateau. The same kind of response applies to birds and their insect prey. Tinbergen (1949) suggested that birds preying upon a variety of insects develop a "searching image" of the favored prey when the overall prey population becomes dense enough to make this profitable.

In all functional responses to prey density there are three universal components: length of time prey are exposed to predator; rate of discovery of prey; and time spent in identifying, capturing, and eating prey. These are logically present in all situations. Some other variables do not operate uniformly in all predator-prey interactions. Hunger, for example! Presumably, searching rate intensifies when a predator is hungry, and searching ceases when it is not. This is generally true, but there are exceptions. Some predators continue to attack and kill even when no longer hungry. Water beetles and weasels are examples. We therefore have a fourth element that usually is present—hunger that leads to intensified search. These four, taken together, are sufficient to explain the arc or curvilinear responses of insect parasites and predators. To them must be added a fifth, in order to explain the sigmoid response curve of vertebrates. Birds develop more complex searching patterns than do insects. The "searching image" mentioned above is a response not only to greater density but also to an intensified stimulus that induces the bird to search selectively for a specific kind of prey. This fact explains the high concentration of birds and the selective feeding on pest species at outbreak numbers. There is indeed a possible sixth component—the response of predators to swarms of prey. From a number of instances it seems clear that very dense aggregations of prey serve to confuse a predator through constant distraction, reducing its efficiency to capture.

But the predator too has its limitations. The functional responses to predator density are reducible to components just as are functional responses to prey. These reactions result largely from exploitation and interference. As parasite and predator populations increase in response to prey density, the chance decreases of discovering prey not already attacked by a given parasite or predator. They are, after all, competing for the same resource; "exploitation" must decrease accordingly. "Interference" simply describes competition between predators themselves. The result of two predators "squabbling" over prey can be only an increase in the time spent in capture and handling. In some animals the act of attacking is not prevented or interfered with but becomes a stimulus for another to do the same. A fish rising after flies stimulates others also to rise even though prey may not be visible. This is not interference but "complementation," a third but subsidiary component which operates only when groups of predators search together. Presumably such a response would be common among mammalian predators that search in loose groups.

Thus far, emphasis has been placed on the effects on prey of either prey or predator numbers. This preoccupation with prey is what is meant by the two *functional* responses—exploitation and interference —I have described. In short, I have described the mechanisms by which predation functions. But there is more to say. Both types of functional responses can affect the numbers of predators present, so that there may be a *numerical* response as well as a functional one. There are many examples illustrating that predators in nature increase as prey density increases. Not only do absolute numbers of predators increase but the number of species of predators may also increase. The rate of intrusion of additional numbers of a single species depends on length of reproductive generation. The rate of intrusion of additional species depends on dispersal rate, which correlates directly with the likelihood of discovering an adequate food source.

Densities of both invertebrate and vertebrate predators respond in these ways. Among vertebrates certain insectivorous birds, for example, are consistently more common during outbreaks of spruce budworm; other birds show no response (Kendeigh, 1947; Morris *et al.*, 1958). Breeding success is rendered more likely at times of high prey density, even though predator density may also be high. Pitelka *et al.* (1955) showed that the numbers, species, and breeding success of avian predators that depend solely on lemmings were related directly to the numbers of this rodent. With these predaceous birds, as with predatory mammals in the Arctic (Elton, 1942), the lag in increase imposed by a reproductive generation longer than that of the prey projects the effects into a subsequent season. Predatory mammals whose cycles of abundance parallel those of prey species always lag noticeably behind both increases and declines of prey numbers. An oscillatory pattern often results. The lynx-hare relationship of Canada and the northern United States is a classical example. The arctic fox responds directly to fluctuations in lemming numbers but with a lag of one generation. When the European pine sawfly is abundant, shrews and deer mice are also abundant. In short, a high prey density increases the total welfare of predators or of predator complexes. Only when numbers of predators increase to very high levels do the functional responses (exploitation and interference) come into play to limit further increase of predator density.

It is obvious that the perceptivity of the predator and the character of the environment will influence the amount of a prey population that a predator can remove. Or, in other terms, the carrying capacity of an environment, as defined earlier, not only describes a balance between predators and prey but interposes a physical element. An area that does not provide adequate food and cover exposes a large prey population to excessive hazard. It has become a tenet in game management, derived principally from Errington's many studies (for example, Errington, 1946), that unusual success of predators indicates a poor habitat more than it does credit to the skill of predators. The prey removed by predation are merely surplus; they would have died from other causes had not predators taken them. This idea is sound enough among vertebrates that feed on vertebrates, but has little meaning among prey species with lesser mobility, such as most insects. Most insects do not achieve the population densities required to produce the density-governed response of which Errington speaks.

One question frequently asked, particularly by wildlife biologists, is: Which components of a prey population are most vulnerable to predation? If all members are not equally subject to the attentions of predators, the significance of partial but selective removal cannot be assessed in a purely mathematical way. Is predation in fact selective; i.e., do predators remove more of the weak, the young, the crippled, the diseased, and the ill-adapted members of a prey population than would be expected by chance?

Though information is lacking to answer this question convincingly, it would seem that invertebrate predators do not normally select particular members of their chosen prey species. The most important feature of invertebrate predation is chance of discovery. Once discovered, entire prey populations seem to be equally vulnerable. An exception would be the many species of parasites that will not oviposit in prey in which eggs have already been deposited. Moreover, the area searched by insect predators and parasites depends very much on the density of prey populations. The successes of biological control confirm that predators and parasites are essentially nonselective. Unfortunately, selectivity is scarcely studied among insects that are not economically important.

It is more difficult to generalize about vertebrates. As Errington (1946) clearly shows, the skills of predators vary greatly, as do also the defenses of vertebrate prey. Moreover, analysis is complicated by territorial behavior and intraspecific intolerance of a degree not occurring in invertebrates. It would seem that warm-blooded vertebrate predators do "select" prey in the sense that the prey individuals that lack secure territories and "cover" become the main targets of predation. These individuals, as Errington (1943) has shown in his study of mink predation on muskrats, are often young or diseased animals. Wolves preying on caribou tend to depend on stragglers and calves (Crisler, 1956; Murie, 1944). In reference to wolf predation on bison, Soper (1941) noted that most individuals killed are subadults ailing in some way or animals in the last stages of senility. Borg (1962) in Sweden has convincingly shown that predators differentially kill diseased roe deer. Of 156 predator-killed deer examined by veterinarians, 54.5 per cent were diseased. Similar examples are many among canid predators, but it should be noted that these animals are especially skilled in the arts of predation (as also are predaceous birds: Rudebeck, 1950, 1951). Even regarding man-eating predators the notion has arisen that children and the more poorly armed are most vulnerable targets. The density of prey and the extent of escape cover are more important determinants of predator success. Nonetheless, in the exposed population, selectivity does operate. Rudebeck (1950), in an excellent study based on many years of field observation, shows that avian predators feeding on birds capture significant numbers of weakened prey. It is a matter of observation common to many species of game birds that young birds are preyed upon more frequently than adults. Fish-eating birds also depend most heavily on younger fishes. In general, then, vertebrate predators do practice some selectivity within prey species and become instruments of "environmental sanitation," enabling established animals to fare better by removing excessive numbers of young and of senile and afflicted animals. The occurrence of selective removal is not doubted; its importance in population control is another question.

We must now look again at one of the two previously posed basic questions—whether predators control the sizes of their prey populations. A most difficult question it is! At this point I am merely continuing the biological accounting, choosing to refrain for the moment from the argument on the necessity of control. (Let us also remember that earlier discussions of food relationships and successional stages in relation to animal representation and abundance are pertinent here.)

A definition is required. "Control" as used here does not mean limiting numbers to those acceptable by man. Rather it refers to the ability of predators to determine the size of a prey population (natural regulation), all other factors being equal. A background thought is also required. Excluding such gross factors of change as geological time, evolutionary process, and cataclysms, one must generally conclude that predator and prey populations exist together in relatively harmonious balance. Excessive predation of sufficient proportions to lower or eliminate prey populations occurs only in specialized situations: the introduction of predators into previously stabilized ecosystems such as oceanic islands; the elimination of alternate food sources for facultative predators; conditions, such as waterfowl refuges, which attempt to induce unnaturally high productivity of desirable species.

Under natural conditions heavy depredation of plants by insects is rare. Defoliation of plants occurs most regularly in crop species and single-type forests. The tendency of leaf-feeding insects to feed up to and beyond their food limits is marked in agricultural plantings; it is rarer in forests (Voûte, 1946), where marked defoliation is neither uniform nor sure. Most insects are kept below their food limits chiefly by the presence of insect parasites and insect predators. The best evidence of this lies in biological control. In the most classic and convincing case, the vedalia beetle, imported to control the mealybug, did so well that the mealybug is now scarce; but so is the beetle. The mealybug is limited by the predator, and the beetle by its food supply. Here as in so many similar examples a greatly simplified situation obtains: one introduced pest and one introduced counter-pest! However, it is possible that two or more parasites or predators might be needed. Acting in concert, they could effectively limit a prey population. Predation by birds on insects is large; astronomical numbers are taken. But there is no confirmation that birds by themselves limit the population numbers of insects in any long-term sense. More significant influences on insect populations are rapid reproductive rates and mortality factors other than bird predation. The important contribution of birds seems to be at low densities rather than high, just as among insectivorous mammals. A relatively higher proportion of prey is taken at lowest densities (contrary to the assumptions of Howard and Fiske, 1911), thus apparently delaying onset of outbreaks and perhaps spacing them more widely.

As with most vertebrates, predators on fishes tend to favor and select certain portions of a prey population. This restrictiveness on the part of a vertebrate predator population stands in contrast to insect predators. The result is that food—or competition for it—not predation, most often limits population numbers of prey. Nonetheless, predation is not without effect. Sizes and growth rates are small in dense fish populations not subject to predation. Moderate to heavy predation where the largest and oldest fish are taken (overfishing, for example) reduces average age but increases growth rate so that younger fish reach larger size more quickly. Heavy predation on fry affects the entire population character. The removal of kingfishers and mergansers in Eastern Canada, for example, clearly led to greater survival rate among young Atlantic salmon, and it is believed that the effect could be noted in greater numbers of older fish (Holling, 1961). Competition for food therefore seems to be at the root of population limitation in fishes; only where predators are skillful and specialized will their effects exceed those of competition. In other terms, these selective predators are obligate feeders on parts of fish populations. When these predators are particularly skillful, predation becomes important in limiting fish numbers.

Birds are common targets of predaceous mammals and carnivorous birds. Here, too, large numbers of birds may be taken as prey, but only in special situations does predation seem to restrict bird populations unduly. In the best study of its kind, Tinbergen (1946) showed that European sparrow hawks took a large percentage of house sparrows during a given year (perhaps 50 per cent). Nonetheless, he could not conclude that the population was importantly impaired from year to year. Moreover, in other areas of England and the Low Countries where sparrow hawks have been assiduously removed, there seems to be no compensating increase in house sparrow populations. A similar conclusion must be reached from the excellent study of Bump *et al.* (1947). In this instance, a variety of avian and mammalian predators harassed grouse. At first sight, combined losses seemed formidably high. Predation on eggs was 39 per cent; on young, 63 per cent; on adults, 50 per cent. Total predation throughout the year took about 80 per cent. To determine the value of predator removal, an energetic campaign was launched to eliminate predators throughout one large study area. The program was particularly effective against the smaller carnivores, but did not eliminate all carnivorous birds. This substantial removal of predators lowered egg loss by half but did not affect predator pressure at other points or times of year. Further, grouse populations did not increase in succeeding years. In this instance, however, the increase in numbers was directed toward enhanced hunter success only a few months away from breeding. It is conjectural whether these increases could be sustained from year to year. Thus, hunting as a form of predation may only remove a surplus that is already doomed. Predation on birds, as Errington's thesis suggests, succeeds only in removing the part of a population not sustainable in any case.

Predation has been studied more intensively in mammals than in any other animal group. Although predation clearly has its effects, it is not established that population limitation is caused primarily by predation. Certain facts are evident. Mammals do not normally devastate their food supply (although in the Arctic this may occur); something other than food must therefore restrict populations. The numbers of mammals taken by predators may at times be very large. Predation on mammals is frequently observed and often dramatic. Population fluctuations, particularly in rodents, are equally dramatic. In some special situations (the celebrated Kaibab deer herd, for example) total predator control has been followed by irruptive increases of herbivores, often leading to overuse of vegetation and subsequent starvation. Irrespective of a bewildering welter of apparently conflicting data and views, it seems likely that predators are not a dominant influence in the control of most mammalian populations. Exceptions would come in those prey species whose reproductive rates are higher than those of their predators. However, predators are not without effect, and, temporarily, they may be responsible for marked reductions in numbers.

Some mammals, like some birds, can be extraordinarily effective hunters. Those mammals with special attributes and a seeming zeal for hunting may be more important factors in control than can currently be documented. As examples, I might note these couplets: foxes–rabbits; fishers–porcupine; mountain lion–deer; and ferret–prairie dog. But even among these skilled hunters, the victims seem most frequently to be the handicapped—the immature, the wanderer, the ill-adapted. The removal of such prey may conceivably result in more vigorous prey populations rather than less.

On a sustained basis predation exerts less persistent control than do habitat limitations, competition for food, and social strife. But a word of caution is necessary. Almost every description one encounters deals with simple relationships: one predator species to one prey species. "Food habits" studies are more often than not tied to single species rated by someone as important. In many instances it is biologically feasible to extrapolate total effects of predation from the study of single pairs of vertebrate predator-prey species. In more species, however, the wider latitudes of feeding habit among birds and mammals dictate against this as a general rule. Very rarely are the effects of predator complexes on pest prey species given proper recognition. Arbitrary selection of single species for consideration normally distorts total evaluation and can lead to biological illusion. Too frequently, the illusion is intentionally sought to justify particular control programs. For example, the excellent study of Fitch and Bentley (see Chapter 17) correctly points out that coyotes did not approach total removal of the annual surplus of ground squirrels in a foothill area of the San Joaquin Valley. This fact has been widely used to justify coyote control campaigns. What is not generally circulated is that coyotes fed widely on other rodent species as well, and that all vertebrate predators, from rattlesnakes through coyotes, acting in concert, did in fact consume most of the year's ground squirrel production. This judgment cannot be extended to other species of rodents, since they were not studied in the same way.

A total complex must be considered. I have wondered why biologists themselves perpetuate a biological illiteracy in which only one-to-one relationships are studied. I can only conclude that the tendency to concentrate on "favored" species in all phases of economic zoology has led to a warped concept of the ecological functions of animals.

Certainly, wider studies in which the total value of predation becomes the core of emphasis are long overdue. Equally overdue are studies that exploit food competition, social strife, and habitat manipulation as approaches to vertebrate pest control. Neither the challenges nor the rewards of such imaginative approaches seem appreciated by the advocates of more toxic poisons, increasingly distributed.

Campaign and Controversy

Of all pest control practices, the killing of predatory animals has elicited the bitterest criticisms. The criticisms vary in intensity and rationale, depending on the kind of predation and the purposes of control. Insects, birds, and mammals figure most prominently in these arguments.

The loss of predaceous and parasitic insects resulting from broadcast nonselective chemicals is now so commonly recognized as not to require verification. Economic entomologists rather uniformly acknowledge the serious nature of this loss and, where possible, seek to avoid it. In practice, avoidance of loss of these important natural controls is sought through appropriate timing of pesticide applications, by continuing search for greater pesticide specificity, by coordination of biological and chemical methods of control, and by substitution, where possible, of introduced species of predators and parasites. The net effect of these efforts is not yet great, and in most agricultural situations major dependence is still placed on chemical control, regardless of consequences to other members of the insect fauna (see Chapter 21). A measure of disagreement—sometimes strong—marks the differing attitudes that entomologists have toward remedying the losses of insect predators and parasites from chemical control practices. All in all, this loss has more general significance than have comparable problems in vertebrate pest control. Nonetheless the arguments among entomologists appear curiously benign to one closely familiar with the violence of controversy characterizing much bird and mammal control practice.

Rodent and predatory-mammal control away from dwellings serves the interests of crop, timber, livestock, and game producers. Less commonly, when, as currently with rabies, a disease transmissible to man appears in wildlife populations, control campaigns may also be justified for reasons of public health. Bird control campaigns (in contrast to local control of nuisance birds) are justified by proponents of control for similar reasons—reduced crop yields, interference with growing operations, and, more rarely, transmission of disease.

Predatory-mammal control has, in the past fifty years, become increasingly sponsored by federal agencies. Local control efforts were and still are based on hunting and trapping. The only really new feature is the use of aircraft, which permits more kills over open country than would otherwise be possible. Hunters and trappers are usually local residents, hired by local producing interests and by governmental agencies at all levels, whose goal is local reduction in numbers of predatory animals. The core of their support has been stockmen, to whom poultrymen have attached more recently. The local hunters and trappers have been aided in making a living by bounties paid for certain species and, under some arrangements, by the sale of fur. The decline in bounty payments in many areas and marked decreases in prices of marketed furs, particularly in the last few years, have made them more dependent on governmental employment than has been traditional.

Hunting and trapping, however, are both expensive to support and local in effect. Beginning just before World War I, a philosophy of area-wide reduction with poison baits began to appear in mammal control operations. The main beneficiaries and proponents were stockmen. Large-scale reductional programs grew greatly during World War I, and have neither declined since nor at any other time expanded so markedly (whatever rationale was used to gain further support). Chief among arguments for support were game production and disease prevention. Neither of these now provides the same degree of justification as characterized the twenties and thirties. Strychnine was the chief chemical used at first, but the German discovery of thallium during World War I gave mammal control a new, highly toxic material not possessing the disadvantages of strychnine. Because thallium lacked both taste and immediate toxic effects, lethal doses were much more likely than with strychnine. For twenty years thallium was the main chemical used in the control of field rodents and predatory mammals. Strychnine had yielded its unintended losses, but thallium produced more (partly because of the chemical's stability and high toxicity, partly because of much expanded control operations, and partly because of the carelessness and indifference of control workers). Opposition to mammal control programs had been voiced against the earlier campaigns with strychnine, but these objections were minuscule compared with those that followed. Beginning in the middle 1920's and continuing for several years, opposition mounted until it became a controversy penetrating state and federal legislatures. In 1930 the annual meeting of the American Society of Mammalogists fully aired the conflict as it existed at that time (see the *Journal of Mammalogy* for 1930 for the several published articles that followed). Resolution of the conflict did not follow; control campaigns did not change character or diminish in scope or intention. In fact, the exigencies of the depression years that followed resulted in more rather than less control, in this instance chiefly as an employment measure. The opposition did not die. Arguments continued through the 1930's. But some changes in control operations did take place. Research on the food habits of animals increased. Workers were assigned to develop new and more selective methods of mammal control. Field biologists were assigned to study how in nature predation actually works and to determine how control measures could be more narrowly tied to the biology of both favored and pest species. More able men began to find their way into the work. "Watchdog" committees were convened to evaluate methods and problems having regional applicability (for example, the Wildlife Administration and Pest Control Relations Committee, convened by the California Senate in 1933, which met until World War II). In retrospect, I must conclude that the controversies and recriminations of the twenty years preceding World War II were not without value. The larger issues were defined; reorganizations of operations and administration were forced to follow the appraisals; a general heightening of competency painfully evolved at all levels. Conflicting values remained to be resolved, then as now. World War II cut short the debate.

Materials production was a yet greater problem in World War II than in World War I. There were more men in uniform, distributed more widely throughout the world, conducting a more technically oriented warfare. Research and development work was diverted to safeguarding food and other materials during the course of their distribution and storage throughout a variety of climates. Because large numbers of the military were forced into contact with new animals and diseases under formidable conditions, great effort in field sanitation and disease-vector control became necessary. Means of controlling rodents and disease became the main preoccupation of research chemists and biologists.

Under the focus of military need, combined efforts can produce astounding results. If one considers mammal control alone, the results of war-time chemical research were indeed impressive. Chief among the discoveries important to our subject was the great effectiveness of sodium fluoroacetate, known more familiarly under its experimental number—1080. It was found to be highly toxic (the most toxic chemical now widely used), soluble in water, and chemically extremely stable; it could be eaten without suspicion by rodents (hence did not produce "bait shyness"); and it did not act so quickly as to interfere with consumption of a lethal dose. Its arrival was hailed as enthusiastically as was the development of DDT. Very soon it was noted that the toxicity of Compound 1080 was not uniform to all animals. Among vertebrate animals, the canids were found to be most sensitive, other carnivores slightly less so, and rodents still less, and birds were several times more resistant than mammals. Toxicologically the differences in toxicity were academic; the chemical remained highly toxic. But combined with biological knowledge and appropriate placement, these differences meant that 1080 could be used selectively to poison only certain kinds of animals. Thus programs became possible in which predatory mammals and rodents could be controlled with presumably minimal hazard to other living things. Compound 1080 has expanded in use since that time (1946–47) and has become the standard chemical in the United States to reduce populations of coyotes and range and forest rodents. It is, of course, used also in urban situations and for more limited control purposes elsewhere. It has been used in large-scale programs in Australia and in New Zealand, where, depending on locality, its targets are deer, opossums, kangaroos, European rabbits, dingoes, and foxes. No other countries use Compound 1080 in the quantities consumed in these three countries. Moreover, in all three, aircraft have been increasingly brought into use to permit broader coverage, better dispersal of poisoned baits, and lower application costs.

In the United States the philosophy of area-wide reduction of pest species was implemented by this chemical so well that little need seemed to remain for other methods. When ground squirrels and coyotes, its chief targets, were widely reduced in number, attention shifted to range and forest rodents. The means were at hand to end pest mammal problems—*if* one accepted only one definition of pest, *if* one agreed that pest control should be conducted by governmental agencies for self-interested production groups, *if* one agreed that public lands were to be treated for the particular advantages of these groups, and *if* one believed that the losses to other wildlife and domestic animals could be written off as scarcely worth mentioning in the face of apparent economic gain.

The use of 1080 early presented problems, mostly foreseen. Hazard assessment was an important phase of its development and its empirical application in the field. On occasion, wildlife losses were great; the methods in use could not guarantee the selectivity anticipated (i.e., that meat-eating animals eat only meat; that grain-eating animals eat only grain). The smaller furbearers became regular unintended targets of 1080-poisoned flesh stations put out for coyotes. The coloration of grain baits was only partially successful in reducing unintended losses of birds. Secondary poisoning, a logical derivative of the chemical's stability, became common, and indeed in some instances was hailed as one more demonstration of Compound 1080's value.

Doubts of the safety of large-scale programs with 1080 came early, and in recent years national opposition has broken out (as has happened in New Zealand).

From my point of view, whatever points might be raised in opposition, the most saddening feature is that the U.S. Fish and Wildlife Service—almost entirely responsible for the development and field use of 1080—has not assumed the remainder of its responsibilities in regard to wildlife. Following the demonstration of field hazard and, latterly, wide opposition to poisoning campaigns of widening scope, the service has not responded (as it did in the late 1930's) by initiating full studies of the effects of its own programs. There have been no continuing appraisals, realignments, and regrouping of able personnel. The response, to this viewer, seems rather to be a tightening within the "control-minded" shell. The aerial treatment of tens of thousands of acres of forest burns in California in 1960 attests to a disturbing direction. I would feel more satisfied with such an operation had the degree of rodent reduction, the rate of population recovery, the actual effects of rodents on reforestation, and the economic gain been studied and announced. Hopefully in the future the pesticide-wildlife investigation program, now figuring prominently in U.S. Fish and Wildlife Service research activity, will examine the hazards of its own control programs. The Department of Agriculture is not the only offender in single-minded poison uses.

To this point I have tried to give a descriptive account of what predation represents biologically and have exercised only such opinion as I believed justified by operations which have mistakenly upset predator-prey relationships. I am inclined to agree with many critics of the specialist that his reporting is often entirely too theoretical and depressingly divorced from practical situations. Let me now make it clear that, however concerned I might be to arrive at the best biological conclusions, I am fully aware that attitudes and judgments on bird and mammal control range well beyond the biology of the matter. Nonetheless the dependence on biological information in the control of noxious vertebrates is much too slim. Where we are ignorant, we seem not to choose the path to learning. When dispute occurs, we do not seem to possess the means or the wisdom to resolve it in mutually acceptable ways. In other terms, the control of warm-blooded animals all too often is neither soundly based on biological fact nor clearly justified by economic need. Superstition, conflicting tradition, self-interest, and absence of fact ensure that dissension shall continue.

No one will challenge the statement that a landowner must protect his property and means of livelihood, or that a human population should be provided with ample foods and fibers for physical and material well-being. I also think that few would doubt that governmental agencies should assist in reducing economic losses where wild species are responsible for depredations. In the control of wild species that figure in disease transmission, public-health agencies must logically be called upon to do what the individual cannot do. We can also accept that the preponderance of wildlife, with its economic, recreational, and aesthetic benefits, deserves protection and encouragement; protective legislation has acknowledged its importance. There is therefore no need to dwell on accepted values.

The doubts about vertebrate control campaigns do not come from these universally accepted measures. The questionable control programs are those which derive in essence from competitive self-interest and which are conditioned by attitudes that too frequently encourage biological and social abuses. The chief criticisms therefore concern who actually gains by the control measure, who must pay for it, and what aftereffects it produces. All three points—gain, pay, and effects—can be and have been translated into social issues well beyond arguments of the self-interested antagonists. Scant as is the biological justification, often fatuous as is the economic rationale, the argument ends with neither biology nor economics. The cultural roots of man are deep. His states of mind, of habit, and of social organization extend the problems of animal control into what is virtually a metaphysical limbo. The control specialist who refuses to consider beyond dictated self-interests for his conclusions lives an illusion. So also does the crop, timber, and livestock producer. Their problems are greater than they believe.

I have already described some general aspects of predation. On biological grounds it was concluded that, among vertebrates, predation alone rarely determines the size of prey populations. Predation reinforces other restricting factors so that prey species will not outstrip the limits of environment. Predation is therefore not without effect. Moreover, particularly where prey are abundant, the less able are taken more frequently than is consistent with chance alone. In special cases where prey populations are low, predation may become a primary cause of population limitation. This fact has particular meaning in prey species, whether or not selectively taken, if the predator breeds more rapidly than the prey (as in the coyote–pronghorn antelope relationship). Whatever the special case, predation tends to restrict prey numbers by slowing population growth rates, and may contain below the limits of environment those prey species that tend to exceed food limits (deer).

Beyond biological considerations of predatory-animal control, four conditioning factors are most important in formulating control measures: the shaping of attitudes; the primacy of economic interest; the favor toward particular game species; and, last in order, the desire to preserve.

Our attitudes toward predators border on biological nonsense. The morality fables of Aesop and the fantasy of Mother Goose, learned in early childhood, hopelessly precondition our views on predation. The wolf in "Little Red Riding Hood" fares no better than the weasel in Burgess' "Danny Meadow Mouse." The picture changes little in the traditional folklore of seasoned hunters and outdoorsmen, and is furthered by the adventure novelists. The real problems of the pioneer were accepted as equally real by his descendants. When the larger predators were eliminated, the smaller were accorded the more fearsome attributes of their predecessors. A host of provocative words— all better expressions of man's feelings than of predators' performances —were and still are used to describe predators: wily, cunning, bloody in fang and claw, sly, beady-eyed, slavering, ravenous, stock-killing, cruel, treacherous. There are others. All have unpleasant implications; several suggest that some animals eat meat, perhaps even human flesh, that on some other occasions these animals skillfully take from us something that we want ourselves, and that their removal will ensure more of this something.

We may be excused the foibles of our heritage. There is some basis in fact for our fears. But we can excuse less well the propagandistic use of these traditional fears by government control agencies. Theirs is a calculated Mother Goose whose announced purpose is to "educate," but whose underlying purpose is to seek unquestioning support for predatory-animal campaigns. Here is one sample, from the *California Livestock News* (Jan. 21, 1958):

Wily coyote and stealthy bobcat numbers were vastly reduced by the day in and day out efforts of hunters and trappers.... San Diego County, always abounding with the coyote menace.... Close by in San Bernardino County, hunter ———— got 111 of the dreaded livestock killers and worriers.

The quotation above is not an isolated one. This is the usual manner of reporting by predator control workers. It is good politics and poor biology, and economically it is supported only through implication.

Where economic liability is clearly demonstrated and the individual owner takes steps to alleviate damage, predator control is easily justified. When problems of alleviation are particularly difficult and affect many landowners, governmental advisory and service agencies can reasonably be called upon. It is beyond this point that acceptance of predator control ceases to be uniform. The common-sense "subsidy" is now replaced with tax-supported agencies that assume the major responsibility for depredation control. Increasingly, livestock growers have come to expect total protection without individual effort. In many areas special levies are assessed against livestock on a per-head basis in partial support of control programs, but these levies rarely cover as much as one-quarter of the cost. Commonly, predator control is directly assisted by county and state funds allocated as bounties in the name of game protection or through transferred funds initially justified as disease control-program allocations. With rare exceptions, the only beneficiary (even this is questionable) is the sheep or poultry producer, who then has control work done for him "free." The control work is, of course, by no means free, and when compounded with subsidies of other sorts, it causes even a tolerant person to pause. The situation is aggravated further by inordinate advantages to the few on public grazing lands, on which totally tax-supported control programs are conducted in the name of "multiple use," while there is concomitantly an abrogation of responsibility for the total welfare of these lands. That economic primacy is a justification cannot be argued; as special privilege, however, it cannot be supported by the general public.

Predator control is no longer supported as a general practice in game production (Latham, 1952; Nagel *et al.*, 1955). The bounty system is now conceded to be conducive to fraud and economic waste and, more important, to contribute no increased game numbers. It is still in force for political reasons in many states and counties, but the weight of accumulated evidence is against its value. There is also a diminution of game-department funds transferred to the predator control activities of other agencies and a decrease in the numbers of professional hunters and trappers maintained by game departments. The wildlife manager has now become convinced that predator control will serve his public only in rare situations.

The preservation of predators is perhaps more difficult to understand than is preservation of many other forms of life. Basically it rests on two attitudes. The first is a mirror image of the Mother Goose conditioning I mentioned. The sight of a grizzly bear at Yellowstone Park, for example, conjures images of daring, primitive times. I vividly recall my excitement when as a boy I first heard a mountain lion cry. A coyote's bark lends excitement and dimension to a still desert evening. Sport hunting of bear, lion, tiger, and so on is based on this atavistic reaction. There is a general desire to keep many of the larger predators. Their sight and sign are links with a past whose raw simplicity has been traditionally romanticized.

The second attitude is more cohesive, less definable, and not so common. This is preservation for its own sake. It matters little what the species—ivory-billed woodpecker, bison, whooping crane, Sumatran rhinoceros, timber wolf, or black-footed ferret. Wherever possible, we seek to avoid exterminating an entire species. Although local extinction of species is an inevitable consequence of human occupation, western culture is characterized by strenuous efforts to safeguard some members of vanishing species in refuges and preserves. The attitude has spread rapidly throughout the world, perhaps as a derivative of European occupation, and is now being accepted as desirable policy by indigenous, nonwestern governments. Taken together, these two reasons for predator preservation are strong arguments against the eradication philosophies espoused by some grower groups and reflected in some public agencies.

The justifications for predatory-animal control and for opposition to control are largely conditioned by the same factors. The justifications are not, of course, of equal weight. The most tangible is economic primacy. The basic assumption of public agencies responsible for predator control is that economic gain results from their programs. A corollary assumption is that the hazards inherent in control programs do not outweigh the benefits derived.

The first assumption rests solely on economics. It would therefore be logical to expect that predator control programing would be argued from a cost-benefit basis. Although a favorable ratio is implied, facts to support it are not furnished. I think this practice is followed, first, because only a small group of the producing citizenry is benefited. Second, because, with benefits being so limited, proper data are not collected to arrive at full and convincing conclusions on the economic advantages of predatory animal control to society as a whole; third, control officials know that producing groups and legislative bodies can be manipulated by exploitation of traditional fears and preconditioning; fourth, a good many control officials are themselves victims of the Mother Goose syndrome I have described.

Logic is not resorted to. Instead a kind of "numbers game" is played, which, in consort with fear and implication, has led successfully to continuous and expanding legislative support of control programs. Here is an illustration from annual reports of the U.S. Fish and Wildlife Service.

The year's catch of predators taken through this cooperation aggregated 94,040, exceeding last year's records by 4,751 and consisted of 88,844 coyotes, 1,360 wolves, 7,189 bobcats and lynxes, 392 bears, and 255 mountain lions. Cooperative campaigns for rodent control involved the treatment of 29,204,282 acres infested with prairie dogs, cotton rats, kangaroo rats, and woodchucks, and of 237,788 premises for the eradication of common brown rats. [1938]

The number of predatory animals taken during the year under this cooperative program was greater than ever before, judging by results in freeing many stockmen of losses by predators. A complete count is impractical under the new methods now used, but the total of those counted was 112,996 (only 4,209 less than last year). The recorded take consisted of 103,982 coyotes, 1,460 wolves, 6,508 bobcats and lynxes, 909 stock-killing bears, and 127 mountain lions. In rodent control operations 12,079,044 acres of infested land were treated for the control of prairie dogs, ground squirrels, pocket gophers, jack rabbits, cotton rats, kangaroo rats, porcupines, and woodchucks. . . . The Service is now in a position to supervise an effective predator- and rodent-control program with an extent of national coverage limited only by the amount of Federal and cooperative funds made available. [1947]

Other control programs, readily solved in sparsely settled areas where highly toxic control programs can be safely used are vastly complicated by increasing human population and the resulting competition for land use. The greater intensity of land use calls for strenuous efforts by the Bureau to keep abreast of multiple demands for animal control in an increasingly complex world. [1959]

Predator control "successes" are not recorded in this last report. These are now left to regional reporters for publication. In November, 1958, for example, as reported in *California Livestock News,* 642 coyotes and 248 bobcats were taken in the state.

Nowhere in these and other reports of predator control organizations are full data presented permitting a dispassionate judgment of the economic benefit of control. I have repeatedly tried to accumulate such data through personal approaches to control administrators. They seem neither to have the needed information nor to be convinced that it is desirable to collect it. Dependence on "number-dropping," innuendo, and advantages extrapolated from predator reduction to stockmen's interests suffices for them. As indicated in earlier sections, the actual costs of both predator and rodent control on rangelands have in some instances been calculated by reliable authorities. Furthermore, it is known that improper use of land leads to increased competition among animals and to a general degradation of total productivity on the land. Finally it is questionable whether there is proper balance among the varied uses to which federal lands may be put.

I submit that numbers of animals taken, acreages treated, personnel assigned, or, for that matter, monies spent do not constitute evidence of damage alleviated. These factors are the simplest to consider but do not reveal the entire picture. The few studies that have been made of stock losses to predation suggest that economic gain in a wide sense is doubtful. Unfortunately the agencies responsible for control programs cannot conduct truly critical investigations of their own operations. Nor could their conclusions presently be accepted wholly. As Durward Allen (1954) eloquently understates the case, "It is well known that popular reports in support of governmental control programs are not distinguished for being critically impartial." Allen's book should be read by those who wish a more complete documentation of the facts and issues involved in animal control.

There is similar dearth of analyses describing whether the hazards of control operations outweigh the benefits. Any benefits which may ensue are far less estimable than is economic gain. In large part, hazards are assessed by control workers themselves. The logical priority of values, which assumes economic primacy, has been interpreted to mean that other values are for practical purposes meaningless, or at best have meaning only as they figure in public and legislative contention. The nature of these hazards is described in other connections elsewhere in this book. What remains to be discussed here is what must be done to ensure proper evaluation of mammal control requirements in the field.

First among these prerequisites for re-evaluation is recognition that research investigations relating to predatory animal control should not end at efforts to kill more effectively. The research workers whose responsibility it is to alleviate animal competition with man should be well-trained biologists, intellectually capable of exploring the full consequences of the presence and controlled removal of animals, and administratively freed to publish, practice, and indeed preach what they, in good conscience, wish. This complex of capabilities is not now commonly possessed by wildlife biologists in governmental employ. Second, a derivative of the first, the representations of administrators to the public and to legislatures should be offered not as fractional segments of problems justified by fractional interests of society, but as total appraisals, judiciously pinpointing where control is needed and where its value is questionable. Third (since I am not such a visionary as to believe that points one and two will be brought about by rational objections and self-imposed reorientation within control agencies), the entire subject of land use in relation to predator control should be fully investigated by Congress, and the recommendations should include that (fourth) the National Academy of Sciences–National Research Council be requested to appraise the problem in all its aspects. On the last point, the recently formed Committee on Pest Control and Wildlife Relationships would be the logical nucleus for such a study. The study should be so amply supported that the many doubts now causing dispute could after investigation be removed by fact, not argument. It is high time to remove both folklore and inordinate special interest from the operation of public predatory-animal control agencies.

17 · FOOD CHAINS AND PESTICIDES: SUBSTITUTION AND COMPETITION

The greatest part of pest control effort is directed toward protection of edible materials. Insects consuming part of a plant are analogous to the coyote attacking lambs and the mosquito seeking a blood meal. Plants are cultivated largely for human food, whether we consume the plant food directly or through an intermediate converter (chiefly livestock). Most animals that transmit disease normally do so in the process of seeking food.

The basic requirement of pest control is understanding the food relationships of animals. An important segment of this understanding is the disturbance of food relationships resulting from pest control measures. Included in food relationships are a welter of topical problems. Each of these problems is important, and sometimes controversial. Rarely are they considered together. The following outline relates these varied aspects of the basic theme.

220

Direct responses to chemical control
  Substitutions
  Competition
  Predator-prey interactions (see Chapter 16)
Indirect effects of chemical control
  Transferred effects—food denial (see Chapter 18)
  Transferred chemicals
    Secondary poisoning (see Chapter 19)
    Delayed expression (see Chapter 20)

Substitutions

The degree to which animals deprived of a favored food source are capable of substituting another is an important consideration in establishing the benefits, hazards, and efficacy of chemical-control procedures. Lability of feeding preference ranges from the obligate feeder (largely restricted to single species, perhaps even to single parts) to the omnivore (limited only by its physical ability to obtain and digest food). The consequences of such variability in animal food habits are many, and are repeatedly touched upon in other contexts throughout this book. Here I wish to present briefly some general features of the food predilections of animals from which practical judgments can be made.

Living plants are the source of food for most animals. Normally, a carnivorous animal is only one or two steps removed from the plant base. Plant-feeding animals constitute the first energy-conversion level on which all predation and parasitism depend. Plants themselves are adapted to live in the presence of animals, and indeed in many instances it is clear that the species composition and appearance of plant communities are shaped by dependent animals. Dependency is, however, mutual. Dispersal of seed, normal successional sequences, and plant form and diversity are in large part maintained by animal components of the biotic community. It has been noted that the angiospermous plants could have achieved their present diversity and distribution only in the presence of mammals rapidly evolving in a parallel way. The diversity of a community leads to the adaptedness of its members.

Few plants can resist the depredations of animals, although many have physical or chemical mechanisms that inhibit some kinds of animal feeders. A great number resist certain types of insects. Toxic plants are one indication of the adaptiveness of plants for resistance to animal attack. Many common pesticides are "botanicals," that is, chemicals derived from plants. Some of these are now synthesized. Pyrethrum extracts, for example, were used insecticidally for centuries, long before the toxic constituents could be chemically identified and synthesized. The clue to their pesticidal value came from simple, unsophisticated observation. Other botanicals are rotenone, ryania, and red squill. Even Compound 1080, the most toxic of rodenticides in use, is a naturally occurring organic salt. Nitrate and selenium poisoning in livestock are reflections of the ability of plants to concentrate toxic materials. The urgency to control *Halogeton* and Klamath weed is based on their toxicity to grazing animals. The planned selection of resistant crop varieties is normally directed at particular disease or insect pests. The base of such resistance may be physical impediment, repellency, or toxicity. More commonly than not, the ability of resistant plants to inhibit certain kinds of pests is clearly seen, but the mechanisms by which it is achieved are poorly known and insufficiently studied. (Martin, 1957b, has reviewed the chemical basis of "antibiosis" in plants.) The means by which plants have become adapted to live in the presence of animals are extremely varied. It would seem highly desirable to intensify research investigation on these means so that full exploitation could be made of them in pest control.

The mobility of animals enables them to search, whether their "prey" be animal or plant. The lesser searching abilities of some animals are countered by high reproductive rates and mechanisms for passive dispersal. Invertebrate animals tend to be less opportunistic in their feeding than vertebrates. Whether herbivore, carnivore, or parasite, arthropods are generally obligated to a particular species or food type. The complexity of the relationships to ensure food specificity is sometimes incredible. An example is the obligate sequences of host dependence among parasites. Moreover, among insects, growth and resting stages often present quite different forms and habits, each of which may be food-specific and completely unlike others. In the species that undergo metamorphosis (which include most insect pests) seasonal adaptations are met by different stages in the life cycle rather than by substitution of other food materials, as is normally the case with vertebrates. The only shift of food preferences in vertebrates comparable to that of the cycling insect occurs at the time when the very young lose parental dependence. From that time forward, opportunism characterizes feeding habits, bounded only by the general nature of the animal. The general conclusion is that vertebrates are more able than invertebrates (particularly insects) to substitute other species and types of food.

The greater dependence of insects upon precise food sources in sequences coordinated with stages in life cycle improves the likelihood of their control. Insect control ideally aims at the most exposed link in a specific food chain, thereby disrupting the entire chain. In control of vertebrates, measures can rarely be so specifically exercised, and it follows that the vertebrate pest is the target of control throughout a greater part of its lifetime. Whether against insects or vertebrates, the approaches to control also determine the attendant hazards. We shall see some of these hazards in the examples that follow.

Food ranges among herbivorous insects vary widely with the species (review by Thorsteinson, 1960). *Monophagy* and *polyphagy* are general terms describing these ranges. The use of these terms does not refute the general rule that, biologically, most insects are narrowly bound at each stage in their life cycles to particular food substrates. Their status, to production entomologists, tends to be judged anthropocentrically—whether the insects feed on one or more than one crop species and whether they do so at intensities sufficient to be called pests. The feeding specificity of insects is attested by the common names of many species: Colorado potato beetle, spruce budworm, wheat stem sawfly, lodgepole pine needle miner, alfalfa weevil, artichoke plume moth, etc. Insects attacking a wider range of crops seem more commonly identified by vulgarized scientific names, morphological tags, or country of origin. Examples are lygus bugs, "white grub," and Japanese beetles.

Chemical control usually aims at a particular pest species but does so with nonselective insecticides capable of killing many other species. If food substitution might occur, it can be obviated by resistant crop varieties, by crop rotation, or by chemical maintenance of faunal sterility. It is rather uncommon for a pest of one crop type to transfer its major attention to another, and in any event what is termed a polyphagous insect is not equally common or destructive to all species on which it might feed. A species might be rated polyphagous ecologically, although it is economically monophagous or nearly so. Chemical control would be infinitely more difficult were this not the case. In other words, the magnitude of insect control problems stems from the great number of narrowly adapted insect species and stages, not from wide-feeding habits or from an ability to substitute freely one food type for another.

Selective-chemical treatment takes advantage of this limitation, and biological control uses it as the rule. Successful weed control with specific insects reduces the controlling insect to low numbers along with its host plant. Similar specificity in predaceous and parasitic insects results in low numbers of both pest and controlling species. In one extreme case, biological control workers have even suggested restoring part of the pest population in order to sustain effective numbers of controlling species. Biological control depends on narrow specificity of feeding (or of oviposition) to ensure reduction of pest populations only. In ecological terms, the food chain is lengthened in relation to a single prey species, but not extended to form additional food-chain relationships.

The fact that food chains for insects are specific and identifiable brings out an important hazard of nonselective chemical-control methods. The inability to confine killing to target species results in a loss of beneficial controlling species. Their slower recovery in numbers, compounded by the reduction of their specific food and their inability to substitute other species as prey, leads to unrestricted growth of surviving pest populations and to diluted success where introductions of predators and parasites are made. These effects are discussed in a more detailed way in a later section (Chapter 21).

Although vertebrates are generally able to substitute foods more freely than most invertebrates, there are many instances wherein this compensatory ability has not been sufficiently labile to prevent food shortage and starvation. The best evidence that this occurs is in vertebrate populations that fluctuate greatly in numbers. Pitelka (1957) emphasizes that lemmings increase to such high numbers that they virtually destroy their food base. Three to four years are required before both plant (sedges) and dependent herbivore can recover. Avian and mammalian predators decline in numbers following a lemming "crash." The simplified tundra ecosystem presents little opportunity for substitutions of food.

In lower latitudes, where alternative plant and animal foods exist, substitutions can be and are made. The ability to substitute is less important among herbivorous vertebrates, whose diet is relatively unspecific, than in carnivorous vertebrates. Nonetheless, herbivores generally take foods that are abundant and change their diet as abundance changes. Deer in California, for example, eat more grass in late fall and early spring than in other seasons. As bitterbrush leaders are browsed off in spring and summer, foraging deer turn to new growth of less favored shrubs and trees. Depredations of elk and deer on orchards and forage crops lead to farmer complaints and attest to the ability of ungulates to shift diet. Jameson (1952) has shown that in the white-footed mouse (*Peromyscus boylei*) diet varies seasonally, depending on preference and availability, and has demonstrated that two species of *Peromyscus* living in the same area do not show parallel lability of food substitution. Different population fluctuations within a single rodent genus are therefore possible. Rodents in cultivated lands also may shift diets opportunistically. Pocket gophers in seeded rangelands, for example, prefer the fleshy vegetation and larger roots of perennial species; but when the favored species are reduced by harvest, seasonal decline, or selective removal by herbivores or herbicides, the gophers turn to the less desirable annual plants. It is abundantly clear that rats and mice will feed on a great variety of stored foods. Graminivorous birds also show opportunistic response to available foods. Their diets may depend variously on seeds, blades of grass, or invertebrates. Diets of waterfowl show similar variation.

Carnivorous vertebrates exhibit a much wider latitude of feeding preferences. For example, the major items of the diet of coyotes are rodents and rabbits. Fruits will be taken in season. Carrion will be eaten when available, and in some instances deer, domestic livestock, and poultry are eaten. Rabbits are preferred, but ready availability surmounts preference in determining the types of food eaten. Abundance and availability also dictate the food of insectivorous birds. There are many recorded examples in which birds have responded to outbreaks of insect pests by concentrating in outbreak areas and feeding exclusively on the pest insects (Cottam and Uhler, 1940). Here are a few: gulls and the Mormon cricket (Henderson, 1931); gulls and a species of grasshopper (York, 1949); warblers in spruce budworm outbreak areas (Dowden *et al.*, 1953); several species of birds in locust outbreaks (Smith and Popov, 1953); and many more are recorded in a recent Russian summary (Poznanin, 1956). Similar concentration of avian predators occurs at times of rodent abundance. The shifting response of carnivorous fishes to available insect hatches is well enough known to fishermen to require no additional comment.

When preferred foods decline in abundance and food predilections are limited, declines through food shortage and starvation must also occur in the numbers of dependent carnivores. This pattern is seen in many food interactions (see Errington, 1946). However, the substitution of other foods often prevents predator starvation or dispersal. These shifts in food are sufficiently common to have given rise to formal recognition in wildlife management of the concept of *buffer species*. Buffer species are alternate foods on which predators depend in times of low densities of preferred prey. Predators can thereby remain in an area and check an increase in prey numbers as it occurs. The literature of game management contains many illustrations (see, for examples, Errington, 1946, and Scott, 1943).

Generally little use has been made of artificial accommodation of conditions to maintain predator densities in pest control programs. In some cases creation of suitable nesting and roosting sites has brought about increases in bird populations. Canadian Forestry workers have exploited a similar idea among insectivorous mammals. Rodent control, on the other hand, has not been distinguished by imaginative attempts to provide alternate prey species or suitable protection for predators. An interesting exception has been offered by D. E. Davis (1957), who conducted a deliberate experiment and analysis of cat–rat relationships. By providing an alternate food (commercial cat food) when needed, he maintained a high predator population (cats) in relation to a prey population (rats around farm buildings) that was kept low by the cats. Rat populations exposed to the artificially sustained predator declined earlier in the fall and did not increase as rapidly as usual in the spring. Experiments of this character are valuable but rare. I would like to see this type of experiment extended to field, range, and forest.

Food shortages resulting from control measures must be common, though few studies have focused on the problem. Cited later in this chapter are several studies dealing with changes in entire populations. The normal measure of food shortage is the percentage reduction of a population. The judgment is, of course, total and terminal. The precise mechanisms by which population reduction is ensured are described only grossly by such terms as food shortage and starvation. There are, nonetheless, several studies suggesting that chemical controls might logically induce a food shortage. The clapper rail, a marsh bird, depends largely on fiddler crabs as food (Oney, 1951). Chemical treatments of marsh areas have repeatedly been shown to result in virtual annihilation of these crustaceans (Rudd and Genelly, 1956; Harrington and Bidlingmayer, 1958). Mollusks and annelid worms, which are relatively untouched by control chemicals, are not readily utilizable by clapper rails. No study has been made that would delimit the extent of population reduction in the treated areas, however probable it may be that reduction occurs. Catastrophic reduction in crayfish numbers and their slow recovery followed a low-level DDT application over Illinois woods and ponds. This loss of crayfish deprived a high raccoon population of a staple item of diet (Couch, 1946). Foxes in England, following severe reduction of rabbit numbers by myxomatosis, turned for food to small rodents, poultry, and game birds. These alternate food sources are not sufficient "buffers," and fox numbers are declining. Fishes are particularly sensitive to contact with chlorinated hydrocarbon insecticides, and near-total fish-kills are common. A host of fish-eating birds are affected by the removal of a prey type for which others cannot be substituted.

I might in summary restate the main points on food substitutions. Generally speaking, arthropods are closely tied to particular food sources; this specificity is an advantage viewed in terms of their control. The same specificity is an ecological hazard if chemical treatment results in the removal of beneficial insects. Vertebrate animals, on the other hand, are only broadly linked to food types, and commonly vary their diets as availability dictates. This opportunism has repeatedly been shown to be valuable in pest outbreaks. Rarer, but not so important, is the fact that vertebrate opportunists take desirable plants and animals as food, usually only when their foods of choice are depleted. The economic liability of vertebrates is countered by the esteem in which vertebrates are held as a group. In contrast, arthropods are not highly regarded.

Competition

Previous sections described competition in its broader meaning. Usually, however, the word is limited to specific kinds of competitive relationships in forestry and in wildlife and range management. Birds and mammals—many of which are game species or otherwise valued —are commonly concerned. In these matters, the problem is to maintain high numbers of favored animals while limiting their depredations on crop or timber species. For example, deer will feed regularly in alfalfa fields, orchards, or regenerating forest plantations. Waterfowl occasionally cause considerable damage to cereal or vegetable crops. Upland game birds or songbirds may damage agricultural crops in a number of ways. The relationship differs from insect depredations on plants chiefly in that most birds and mammals are accorded particular esteem. Moreover, applied biologists who deal with bird and mammal depredations differ markedly from other pest control workers by reason of their values, training, and professional alliances. The problems of these workers are many and difficult.

Equally challenging is the question of herbivore competition on grasslands. I wish to dwell on it here because I consider it one of the most misunderstood phases of pest control. Far too much ignorance, misapplied biology, and contention surround the question.

Let us observe the drama of the grasslands. Our stage is the natural grasslands throughout the world. To these many millions of acres must be added the lesser millions that have been put into pasture or that can be converted into partial grasslands. In California, for example, vigorous programs are under way to "improve" some 20 million acres of chaparral or brushland by converting it into grassy cover. The problem of herbivore use of forage grasses is compounded by the ownership of land. The United States Department of Agriculture and Department of the Interior control some 400 million acres of land, chiefly in the western states. Almost half this total is open to livestock grazing; management of all is intended to be guided by a concept of widest use but multiple purposes are not commonly accepted by private owners of grazing land. Similar conflicts of purpose are found, but with no less intensity, in lesser acreages in parks, refuges, and military reservations in all parts of the country.

The heroes and heroines of the drama are the ungulates who convert forage grasses into useful animal products. These products are chiefly meat and animal fiber from domestic animals, primarily sheep and cattle. The maintenance of wild ungulates, chiefly the cervids, deer and elk, derives from their importance in hunting, recreation, scientific investigation, and species preservation.

The presumed villains of the piece are insects, rabbits, rodents, and occasionally deer, which divert forage into unusable channels or which, as often charged, contribute to the deterioration of grazing lands. The marginal villains are the predatory mammals that complicate the plot.

An analysis of competition, thus limited, begins at its simplest with the question whether the various herbivores actually do compete with one another. The derivative questions are many and varied, often molded by the special interests of particular groups. Here are a few questions to illustrate the spectrum of problems:

Are high rodent populations a cause or an effect of range deterioration?

Can grasshopper outbreaks be linked to overgrazing and other land abuse?

Do cattle, sheep, and deer eat the same foods, thus competing for forage?

Why is it that once-good grazing lands can become unproductive, rapidly eroding semideserts?

How can one distinguish the deteriorating effects of recurring drought from those produced by herbivorous animals?

Do native herbivores contribute to the formation of good grasslands (through soil formation and stabilization, seed dispersal, fertilization, plant stability, and so on)?

Can abused rangelands be restored to former productivity?

Do predatory mammals keep range rodent populations in check?

Are range insect and rodent control programs biologically or economically sound?

To what extent should tax funds be used to maintain commercial grazing interests on public lands?

The questioning could go on. The answering is far more difficult and involves social, economic, biological, and aesthetic valuations. I certainly will not attempt conclusive answers, nor do I believe that anyone can now offer generally acceptable solutions for the many problems faced in grassland management. For background, a reader should refer to the summary treatments of Dasmann (1959), Darling (1956), E. H. Graham (1944), Koford (1958), and Sampson (1952). The following brief account centers on the concepts and practices particularly important in range pest control.

Much range damage comes about because of a failure of range owners to realize that each area of range has a limited *carrying capacity*. Carrying capacity can be defined as the number of grazing animals that can be maintained in good condition without damage to soil or range forage. Grasses can withstand only a limited amount of pressure. Perennial grasses produce a surplus of foliage, which can be safely removed. The foliage of annual grasses can theoretically be totally removed after seed formation. In both instances, however, a portion of foliage must remain to provide soil protection and permit seed germination. The species of grasses are not uniformly resistant to grazing pressure. Some are hardy; others, easily destroyed. The species composition frequently changes in response to overuse and to the introduction of non-native species. Sustained abuse has the effect of keeping the plant formation at early successional stages. Under these conditions, introduced species dominate more easily since they are not members of the normal successional sequence. By this process, once-climax grasslands may be altered into quite different, though relatively stable, formations. These disturbance climaxes are now common substitutes for original grasslands. Even the complete removal of livestock does not lead to re-establishment of original climaxes. In California, for example, rangelands have gradually become stable annual-grass communities unlike the original cover. Range-management efforts now concentrate on making the most of these new community compositions or on attempts to restore, through introduction, grasslands approximating the original character. Livestock are, of course, also introduced. Introduced plants and grazing animals have thus changed climax relationships. Reduction in numbers of large native herbivores has been more than matched by "introduced" numbers of domestic livestock. Changes in grassland composition, particularly those associated with overuse by domestic animals, have resulted in shifts of relations among smaller native herbivores, sometimes to damaging proportions (see Chapter 15). In some countries exotic introductions have seriously damaged productive grazing lands, beyond abuse of lands by overgrazing livestock. Examples are *Opuntia* cactus and the European rabbit in Australia.

The extent to which native rodents compete with livestock for forage has particular economic significance. It is clear that indigenous rodents often increase in number in the presence of grazing. Jack rabbits in Arizona, for example, were most common in moderately grazed lands, less abundant in lightly grazed lands, and quite rare in excellent rangeland from which livestock had been excluded as well as in range which had been totally depleted by serious overgrazing (Taylor *et al.*, 1935). In the days before human settlement the habitations of prairie dogs depended on localized damage by bison to the original short-grass prairie. The conversion of original prairie to cattle grazing, widely attended by overuse, furthered both extension of range and increase in numbers of prairie dogs (Koford, 1958; R. E. Smith, 1958).

On desert rangeland in New Mexico, jack rabbits and several species of rodents were more common in heavily grazed areas (Norris, 1950). Ground squirrels and kangaroo rats thrived best in areas of moderate to intense grazing pressure, according to Fitch and Bentley (1949). The pocket gopher was more abundant and destructive in mountain meadows that heavy grazing had caused to revert to early successional stages characterized by weedy forbs and grasses (Moore and Reid, 1951).

It is also clear from the above accounts that increased rodent numbers following intensive use of rangeland do pose problems for the livestock grower. Considered over a broad area, rodents on well-managed grassland are neither abundant nor particularly competitive. On heavily grazed areas a variety of animal "weeds" become abundant and do in fact remove forage that presumably could have been used by livestock. The restoration of poor range to good productive status is often hindered by rodents. In the absence of population control, a relatively stable, early successional (hence poor) stage can be maintained. On very poor range, rodents and rabbits can contribute to further deterioration, thus ultimately leading to their own destruction.

In no case has it been shown that the primary cause of range deterioration is range rodents. On the contrary, the presence of introduced livestock, often out of all proportion to the carrying capacity of the range, initiates the deterioration which in turn contributes to the prosperity of range rodents and rabbits. The large native ungulates that cattle replaced produced only local responses of this kind. In contrast to cattle, they were nomadic, they fed selectively, and only in a few places did they exert continuous grazing pressure.

Acknowledging that range rabbits and rodents do in some instances compete for forage and that in most instances the presence of rodents interferes with the restoration of desirable range composition, we are left with an important practical question. Is it economically feasible to reduce (usually with poisoned bait) rodent and rabbit populations on rangeland? Several considerations go into an answer. It must be established that competition exists in significant proportions. Ignorance and exaggeration complicate this judgment. Normally, maximal or potential damage is reported as if it were uniform and commonplace. In an excellent study, Fitch and Bentley (1949) followed small plot enclosures stocked at unusually high densities of single species of rodents for several years and showed that damage to forage was significant. Although the authors carefully qualified their results and warned against undue extrapolations, their figures are widely circulated as evidence of serious damage to rangeland. Under more representative conditions, Norris (1950) concluded from eight years of study that rodent control had no purpose in well-preserved rangeland, and that the continuous poisoning necessary to increase forage yields on poorer land could not be economically justified. Vorhies (1936) calculated that to remove the number of black-tailed jack rabbits equivalent to one cow would cost $47.36—at a time when livestock brought perhaps $25 a head. Often overlooked—and in need of further serious study—are the beneficial contributions of range animals. These may be greater than generally believed. For example, Bond (1945), in a thoughtful review, has suggested that rabbits and some rodents may assist the recovery of range. By exerting differential pressure on preferred food plants typical of disturbed areas, they may actually speed the return of climax grasses. Additional beneficial values might include fertilization, seed storage, soil aeration, soil turnover, and water percolation (see, for example, Formosov and Kodachova, 1961).

In special situations, where range restoration has value beyond grazing or where conversion of cover type results in relatively high productivity, control may be both required and economically feasible. Nonetheless, the blame for high populations of rodents, low carrying capacities, and deteriorated ranges should be placed squarely where it belongs. This sickness of land is due primarily to overstocking; rodents and rabbits are secondary factors. I can do no better than to echo the thought of the eminent range ecologist, W. P. Taylor: The best control for range rodents is a three-strand barbed-wire fence.

Advocacy of broad-scale range rodent control continues and is increasing. There is as yet no effective method of controlling established range rodents other than poisoned grain baits. Currently, the chemical chosen is Compound 1080. Hand application can be economically justified only where labor cost is low. In view of prohibitive labor cost the trend is toward aerial distribution over large acreages (Howard *et al.*, 1956). In more restricted situations toxaphene, dieldrin, and endrin may be applied as mammal poisons. Advocates are most commonly found in colleges of agriculture and in governmental agencies, and are guided, single-mindedly and without ecological awareness, by livestock grazing interests. The furtherance of these programs is abetted by strong livestock organizations and by legislatures, particularly in the western United States, dominated by livestock interests. Considering that control by government agencies costs the livestock grower almost nothing, this is not surprising. Support for the control of range rodents seems readily available; money for investigations and appraisal seems strangely scarce.

Herbivores other than rodents also compete for range forage. Native ungulates are frequently charged with serious competition with sheep and cattle. The charge is partially correct, but overstated. Sheep, cattle, and deer do at certain seasons eat the same foods. William Longhurst (personal communication), of the University of California's Hopland Field Station, concluded after many years of study that sheep and deer do not compete as much as is generally believed, deer being seasonally far more selective feeders. Buechner (1950) has shown that pronghorn antelope in Texas do better on rangeland grazed at moderate, rather than light, intensity by cattle. Deer also seem to increase following some changes in range use. The fact of competition is well established; its degree is probably rather consistently overdrawn. In this country the remaining native ungulates are carefully managed as far as crop damage is concerned, and in no instance is poison an instrument of management. Contact between crops and deer is best avoided by fencing and reduction in numbers achieved by shooting. The commonly accepted desire to maintain native ungulates at maximum convenient numbers softens their competitive position in relation to livestock.

The competition between livestock and insects (grasshoppers) can also be severe but is more apt to be occasional. Much the same conditions favoring increase of rodents bear on irruptive increases of grasshoppers on rangeland, but in addition these insects respond to aridity of climate. Their occasional increase to devastating numbers, coupled with changes from sedentary to migratory habits, has been noted throughout recorded history. (In common use, "grasshopper" refers to insects at low density in sedentary populations; "locust," to migrating swarms.) Outbreak levels normally develop at focal points that widen over successive years. Migrating swarms may travel hundreds of miles, frequently causing catastrophic damage to field crops and native vegetation en route.

Throughout the world grasshoppers (or locusts) thrive best in open, semiarid country with sparse vegetation. Oviposition and nymphal development are favored under these conditions. Environmental factors that produce these conditions are of two kinds—periods of drought, which cause reduction of vegetative cover and encourage drying out and cracking of land, and practices of grazing and cultivation that do the same.

Recurring droughts are strongly linked to periodic outbreaks of grasshoppers (Gunn, 1960). Recall that these insects are always present in low numbers in those areas where outbreaks occur. Their sudden increase is the natural product of induced biotic simplification caused by climatic extremes. In the United States the main area where excessive numbers occur is in the northern Great Plains states. The "Dust Bowl" produced both migratory grasshoppers and migratory workers. Grasshoppers, however, occur widely throughout the United States. Sedentary populations responsible for crop losses find favorable sites for oviposition along roadways, ditchbanks, and other disturbed but uncultivated fringes adjacent to croplands. The simple expedient of planting uniform grassy covers reduces their numbers by discouraging reproduction.

The continuous intensive use of rangeland, always subject to prolonged periods of low rainfall, contributes also to production of a favored environment for grasshoppers. There can be little question that overgrazing by livestock leads to overgrazing by grasshoppers. Nor does it seem questionable that, for the reasons mentioned, the periods between damaging levels of grasshopper numbers are shortening.

Several observers have noted the direct relation between livestock and grasshopper numbers. Grasshoppers in Kansas were few in ungrazed lands compared to the numbers in immediately adjacent heavily grazed areas (R. E. Smith, 1958). Weese (in Taylor *et al.*, 1935) noted that orthopterans (grasshoppers) in Oklahoma were eight times as abundant on overgrazed lands as on conservatively grazed lands. The connection between range depletion and outbreak of noxious grasshoppers in British Columbia has been clearly traced by Treherne and Bucknell (1924). They report in sum, "Judicious management of cattle within selected grazing limits is the keynote of success in grasshopper control on the range." On other continents too, the densities of grasshoppers of some species can be related to the presence of livestock. In North Africa, for example, the Moroccan locust thrives on the denuded stony deserts heavily grazed by sheep and goats (Uvarov, 1957). In eastern Australia the locust plagues are to a large extent man-made, and have become greater since the country was settled by Europeans

(Clark, 1947). Continuing outbreak centers have come about because of clearing and because of the excessive use of grasses by sheep, cattle, and rabbits. Certain kinds of grasses preferred by locusts for food and shelter were carried to Australia on the wool of sheep, and have further enhanced the welfare of these insects. It would seem from widely separated reports that grasshopper numbers are significantly associated with intensive use of grazing land. Perhaps here, too, the three-strand barbed-wire fence can be called upon to control these animal "weeds."

In practice, however, grasshoppers and locusts are now controlled by broadcast chlorinated hydrocarbon chemicals covering millions of acres annually. Ecological concepts of control, considered before World War II, are now scarcely recognized, much less put into practice. On our western rangelands, grasshopper control programs have for the last several years depended on a 2-ounce-per-acre aldrin spray. Many millions of acres have been covered, some repeatedly, in this way. The biological causes of outbreak have not been corrected, nor is there much investigation of the biological effects of this spray regime. As in range rodent control, the specialist operates single-mindedly for temporary alleviation of economically damaging numbers, unconcerned with the potential of ecological management to solve the problem permanently.

Basically, the need for pest control on rangeland as now conceived is a consequence of overuse of land. The semiarid lands of the world cannot sustain the sedentary pastoralism that we have come to think of as desirable. Whether in Israel or New Mexico, the results of continuous intensive grazing are painfully apparent. The lessons seem not to have been learned well. What I have been describing as "competition" is only a symptom of land abuse. Only lessening of this abuse will prevent further impoverishment and deterioration of rangeland. Present range rodent and insect control practices are, at best, palliatives.

18 · FOOD CHAINS AND PESTICIDES: TRANSFERRED EFFECTS OF CHEMICAL CONTROL

Any animal reflects in some way any change in the food supply to which it is adapted, and we depend on this fact in a variety of cultural and biological control techniques. It should surprise no one that the same kind of response occurs in all species of animals adapted to or favoring particular types of food. However, the fact seems not widely appreciated that any means of control that even temporarily removes large segments of food populations must inevitably have its consequences on the feeders. Chemical control, particularly, while removing most of a pest population also effects reductions in populations of other species in the area under treatment. Food shortages must follow, which, if extreme, lead to debility, starvation, or emigration. The responses of an individual animal thus affected may be immediate, but not uncommonly they are delayed for considerable periods and expressed in ways that at first sight may not seem associated with the initial cause. In this section I wish to emphasize that the problems of

236

food shortage induced by chemical agents are not imagined or theoretical, and that they are frequently disguised and delayed. Direct toxic effects of chemicals on the feeding species are not now under discussion.

Earlier I described how many birds and mammals were attracted to concentrations of prey species. Conversely, a prey population suddenly depleted by broadcast chemical treatment cannot sustain its feeders. If sufficiently mobile, the feeders will leave a treated area. This movement has been repeatedly observed among insectivorous birds (see Rudd and Genelly, 1956). For example, there was no mortality apparent among birds on Montana rangeland following a single application of Sevin at one pound per acre for grasshopper control. Yet, within two weeks of spraying, bird counts on the study area dropped from 173 to 30 (DeWitt and George, 1960). In another instance, again with no visible mortality, all insectivorous birds, except woodpeckers, left within three days when an Illinois bottomland was treated in midsummer with DDT at one-half pound per acre (Couch, 1946). Not all birds are able to leave an area under treatment, however, and all are restricted during the breeding season. In birds as in other animals there has not been sufficient study of the immediate reactions of sedentary species to a food supply suddenly reduced. In many reports in which declines in numbers are noted after a few days, the decline is generally ascribed to direct toxic action of the chemical applied. I suspect that food shortage, rather than direct chemical toxicity, is often responsible for observed mortality. Effects of this kind are more apt to occur among weakened, rapidly growing, or very young animals, and will have ever greater significance as the size of area under treatment increases.

The effects of food shortage are more clearly seen over longer periods. For convenience, these may be referred to as seasonally delayed effects, although the effects may be felt for several subsequent seasons. In all known instances, sedentary species have shown the main responses. In one sense they are captives of an ecosystem—species strongly dependent on a single type of food or physical environment. Normally, chemical reduction of food sources has little immediate impact during the more productive seasons. When the reduced food populations are carried over to adverse seasons and are rendered less available by seasonal inactivity, the combined reductions—chemical and natural—can produce wholesale starvation. This sequence has been observed in a number of cases and is one of the more serious problems of chemical use (see Rudd and Genelly, 1956: pp. 75–76).

One of the best illustrations comes from an extensive die-off of fish in the Yellowstone River drainage following a spruce budworm treatment in midsummer with DDT (Cope and Springer, 1958). Three months after spraying (in October), large numbers of trout, whitefish, and suckers, including many young of the year, were found dead along more than a 100-mile stretch of the river. Reduction in numbers of aquatic invertebrates was great. The loss of food, perhaps not serious while additional insect hatches were appearing in the summer months, became critical when cold weather reduced hatches to zero. In this instance, as in many others, the mobilization of fat reserves containing low levels of DDT may have contributed to the mortality. There is abundant evidence that this happens under starvation conditions in the laboratory. To separate analytically all influencing factors under field conditions is virtually impossible.

Additional studies in Montana and Wyoming have confirmed that reductions in numbers of aquatic invertebrates, and subsequently vertebrates, are great following nearby pest control treatments (Cope, 1960, 1961; Graham and Scott, 1959). Although few fish succumbed to the standard DDT treatment immediately after spraying, late-season die-offs did again occur. In this instance, DDT residues were found in all fish, but the disparity of amounts, combined with many ecological variables, did not warrant direct correlation of DDT and delayed fish mortality. In two intensively studied streams game fish populations declined seriously. Average decline was 75 per cent the second year after treatment. Food abundance, type of food organisms, and tissue contamination all seem to contribute to these declines.

A similar pattern emerged from forest-spraying operations on the Miramichi River in New Brunswick (Kerswill, 1958). There the target was, again, the spruce budworm, but in contrast to most forest spraying in the United States, the application rate was only 0.5 pound of DDT per acre. Mortality was heavy among young Atlantic salmon (to 91 per cent) and aquatic invertebrates (100 per cent for some types). Late-season starvation further lowered numbers among the survivors of chemical treatment. Adult salmon in the stream at the time of spraying were not affected. The hazard here lies in mortality of young, which would go to sea for periods of one to four years before returning to fresh water to spawn. Efforts are now being made to restock populations of the larger aquatic insects in order to provide ample food of appropriate size for older age groups of salmon. The combined effects of direct kill and starvation have already significantly reduced the sizes of salmon runs. In 1962 the run was less than 20 per cent of normal (Kerswill, personal communication).

Delayed effects of this kind are not limited to fish and fish-food organisms, but are most pronounced in them. The edges of waters treated with insecticides show catastrophic reductions in invertebrate numbers. Crustaceans are particularly susceptible to poisoning. I have previously described heavy crayfish losses in fresh waters, and noted the heavy dependence of raccoons on these crustaceans. The unusually high mortality of tidal-marsh crustaceans, along with the probable food shortages that face birds and mammals feeding on them, has also been described. In these instances studies have not been carried far enough to determine the exact effects that continued low supplies of food have on their dependent populations.

Chemically induced food deprivations among terrestrial faunas similarly occur. One study indirectly suggests that food limitations may have important long-range effects on bird populations. A scrub forest in Maryland received five annual applications of DDT at the rate of two pounds per acre. An overall decline of 26 per cent occurred in the local populations of birds in a four-year period (Robbins *et al.*, 1951). In the studies relating to fire ant control it is very clear that reductions have occurred in many animal types (DeWitt and George, 1960). Insects and all vertebrate groups have responded by sustained declines in numbers. In the presence of a stable chemical applied at relatively high rates it becomes difficult to quantify the separate contributions of chemical toxicity and of induced food loss with observed declines. Both contribute, but the degree to which each plays its role is conjectural. Probably one merely reinforces the other.

The significance of food reduction may not become clear in a single year's study. Most investigations are limited to single seasons at best and, more frequently, to immediate "before-and-after" checks. This type of study has value in some situations, but cannot be extrapolated to judge effects of large-scale applications of residual chemicals or even long-range effects of less stable chemicals. Logic and current biological knowledge are sufficient to suspect serious derangements of fauna, although intensive studies extended over several years may be lacking. The emphasis on the direct-killing ability of chemicals is partly a product of preconditioned attitudes similar to the Mother Goose indoctrination I described earlier. But starvation and death indirectly derived from chemical treatment connote no prettier picture, whether or not visible to all. Delayed losses and faunal derangements require much more study. They are, I believe, vastly underestimated.

19 · FOOD CHAINS AND PESTICIDES: TRANSFERRED CHEMICALS AND SECONDARY POISONING

Death of an animal from eating a poisoned animal is a case of secondary poisoning. In commonest use, a simple one-to-one relationship between target species and the animal that feeds upon it is intended. The contamination of tissue with pesticides from a food source may or may not qualify as secondary poisoning. When such contamination kills at a much later date (as I have described), then it qualifies as secondary poisoning, whether or not food shortages, growth demands, or climatic stresses call the stored poison into play. For convenience, however, this phenomenon is better described as *delayed toxicity*. Similarly the transferral of chemicals along food chains in a way that kills only the terminal member is a secondary poisoning which I have called *delayed expression;* it is described in the next chapter. For simplification of discussion here, I have restricted the definition to that kind of relationship in which both animals die. Sublethal poisoning in the second animal must certainly occur in nature, but I have no data confirming this.

Inherent toxicity or ability to poison animals beyond the initial target is not possessed by many poisons, but several poisons in commonest use do have that character. While not all methods of applying chemicals lead to likely transfer to a second species, enough do to make secondary poisoning a serious problem. The general requirements for secondary poisoning are use of a highly toxic chemical; relative stability of the chemical in living tissue; a food relationship between "target" species and feeder; a "bait" substance (usually) that not only attracts target species but concentrates the chemical.

To my knowledge, secondary poisoning in the field has been demonstrated among insecticides only with DDT, aldrin, dieldrin, and Systox and among rodenticides with zinc phosphide, strychnine, thallium, and Compound 1080. Other chemicals do indeed have the ability to poison secondarily, but such occurrence has not been observed in nature. Generally speaking, only the mammal poisons consistently and predictably pose the hazard. Because of the nature of their targets they must be highly toxic and stable, and presented on attractive foods (baits).

Secondary poisoning has figured in these major animal couplets:

Systox-poisoned insects and insect predators
DDT-poisoned insects and trout
DDT-poisoned insects and amphibians or reptiles
DDT-poisoned fish and aquatic reptiles
DDT-poisoned insects and both nestling and adult birds
1080-poisoned rodents or carnivores and carrion-feeding birds
Thallium-poisoned rodents and raptorial birds
Thallium-poisoned rodents and small carnivores
1080-poisoned rodents and coyotes or foxes
1080-poisoned rodents and domestic animals (dogs, cats, pigs)
Warfarin-poisoned rodents and domestic animals
Aldrin- and dieldrin-poisoned wood pigeons and carnivores (including birds of prey)

The control methods that presently give rise to potentially serious secondary poisoning are: aerial spraying over forests, particularly where there are watercourses; poisoned grain baits in rodent control; poisoned flesh baits (stations) for carnivore control; toxic dressings on planted seed in cropland; systemic insecticides used on field crops.

Rarely has the question of secondary poisoning of insects been examined. However, when the systemic insecticides were offered as selective agents killing sucking insects and sparing their predators and parasites, some workers questioned whether the chemicals were quite so harmless to beneficial insects as suggested. Until about 1954 literature on the effects of systemic insecticides on beneficial insects was confusing and inconclusive. It appeared from circumstantial evidence that secondary poisoning did occur among some species of predators exposed through their prey to certain kinds of chemicals. Ahmed *et al.* (1954) then clearly demonstrated that the larvae of several species of syrphid flies were highly susceptible to cotton aphids that had been killed by Systox. Some larvae of coccinellid beetles were also highly susceptible. In no case did the adults appear to react to the presence of Systox in their prey. Since his return to Egypt, Ahmed (1955) has confirmed this accounting with other species, with other systemic chemicals, and in a different environment. Even with these results, it is better to report that the extent or severity of secondary poisoning among insects is not known. Apparently it has not been observed to occur following applications of classes of chemicals other than "systemics."

After many years of experience with forest-spraying regimes, it now seems reasonable to conclude that secondary poisoning of fish occurs consistently but is not of major proportions. Insect victims of forest-insect control are regularly fed upon by fish. Susceptible fish are primarily those that feed at the surface on adult insects knocked down from the forest cover. Presumably, these insects carry more DDT on their bodies than is required to kill them. Although aquatic nymphs and larvae are themselves fatally affected by DDT, they do not have a significant surface contamination. They are not "poisoned bait," as are insects killed in air. Moreover, surface-feeding fishes must frequently break the oil "slick" that accompanies DDT spraying over forests. This twofold contamination appears to be responsible for such fish mortality. Langford (1949), for example, noted that speckled trout (*Salvelinus*) were particularly susceptible to secondary poisoning. This observation has been confirmed since in other areas.

Poisoning is aggravated when these conditions are met and young, weakened, or partially starved fish feed on poisoned insects (Hoffmann and Surber, 1949a). Here too, aerial contamination of insects is required before toxicity appears. Bass, bluegill, and crappie were all killed in this fashion, with bass the most susceptible.

The same set of conditioning factors must explain the few reports of secondary poisoning of frogs, turtles, and water snakes. Herald

(1949) attributed deaths among two species of turtles to consumption of DDT-killed bluegill. Of four snakes that ate frogs poisoned by eating DDT-killed insects, one died (Logier, 1949)—an example of tertiary poisoning. There are other instances in which amphibians and reptiles have been the apparent victims of secondary poisoning, but it is probable that their habit of surface feeding contributed to mortality.

Mammals do not seem to feed extensively on DDT-killed insects. Birds frequently do, and secondary poisoning of small insectivorous birds has been reported in a few instances. No doubt, tissue contamination commonly follows ingestion of poisoned insects, but secondary poisoning appears to be rare among well-fed adult birds. In nestling birds, and perhaps in fledglings growing rapidly, secondary poisoning is more likely and has been described in several instances. George and Mitchell (1947), for example, placed nestlings on a minimal diet consisting entirely of DDT-killed insects, and significant mortality resulted. In other instances, no apparent effect of treatment under field conditions at usual application rates could be observed. I suspect that here, also, special conditions must be met. Higher application rates are probably required to produce a general hazard to young birds. Because nesting habits of birds vary greatly, the likelihood of direct contamination must also vary. Among insectivorous species the incidence of secondary poisoning is probably greater in those species with exposed nests.

Treating seed with fungicides and insecticides is a rapidly expanding practice. Birds that eat broadcast seeds treated with mercurial compounds frequently die. In Finland, Sweden, England, and the United States noticeable losses of birds have followed field applications of seeds treated with mercury compounds, aldrin, and dieldrin. Unquestionably, tissue contamination is present among the survivors. What is less clear is the likelihood of secondary poisoning from these seed "dressings." Falconers in England suspect that carnivorous birds have died from eating wood pigeons that had been feeding on dieldrin-coated seed. Recent information from England (Ministry Agr., Fish., and Food, 1961) confirmed that losses of seed-eating birds had been unusually heavy in the three preceding years, and that, moreover, secondary poisoning was likely among carnivorous mammals feeding upon them. Foxes and badgers have died in some numbers in areas where primary poisoning of birds has been severe. Both field observations and chemical analyses support the suspicion that secondary poisoning is common.

Rapidly acting secondary poisoning from insecticides seems unlikely among warm-blooded animals, according to published information. The fact of tissue contamination is another thing. I think it possible that secondary poisoning might operate in the presence of slowly accumulating pesticides (such as chlorinated hydrocarbons) in a manner not unlike the "delayed expression" to be described in the following chapter: All the conditions would be met for the restricted definition of secondary poisoning previously offered, except that the time interval would be extended. Although there is now no confirming evidence, I suspect that this variety of secondary poisoning might be important; and if it exists, it is potentially very serious. This is true because seed treatment and crop spraying are regular practices, at least once per year, on croplands—in contrast to the occasional timing of forest sprayings. The gradual build-up of pesticides to toxic levels in tissues requires a continuous source of contamination.

Large-scale rodent control enterprises are at present the most serious source of rapidly acting secondary poisoning. In this country, range and field rodent campaigns have been directed largely against ground squirrels, prairie dogs, pocket gophers, and voles. The smaller range rodents have recently become targets for extensive treatment. Compound 1080 is currently the chemical of choice in this country, as it is in Australia and New Zealand where larger herbivores (deer, opossums, and European rabbits) are subjects of intensive control. Thallium, once widely used in the United States, has been almost totally replaced by Compound 1080. Thallium is still commonly used in Europe. Flesh baits used in predatory-mammal control are responsible to a lesser extent for the secondary loss of carrion-feeding birds and mammals.

Since a previous study has reviewed secondary poisoning from mammal control campaigns in some detail (Rudd and Genelly, 1956), I shall not discuss it extensively here. Furthermore, the methods and attitudes of mammal control have already been described (see Chapter 16). Here, then, I shall indicate the main groups secondarily affected, and give some estimates of probable loss.

Birds of prey have been killed by ingesting poisoned rodents. Some species seem more resistant than others. Eagles and the small hawks,

I would judge, are more sensitive than the broad-winged hawks. Eagles, for example, have been killed by prairie dog control campaigns (Arnold, 1954). In Europe, kestrels have been killed in some numbers by thallium-poisoned mice. Owls have been killed in the same way (Steiniger, 1952). Gulls, where they concentrate at sites of rodent abundance, have been killed following control efforts. Apparently, not all members of an exposed bird population are equally susceptible, but the reasons for this differential susceptibility are not known. The extent of prior starvation is possibly critical.

Carrion-feeding birds preferentially feed on entrails and gastric contents rather than entire carcasses. Eagles, vultures, ravens, crows, and magpies have been commonly observed feeding on poisoned rodents and carnivores in this way. Only the vultures are spared the ill effects of poisoning, because of their ability to regurgitate easily. Poisons are concentrated in the viscera of their victims. This unassimilated portion may be sufficient to kill a carrion-feeder. Moreover, most target animals in the field probably consume more than a lethal dose. Herein lies a problem. Extrapolations of laboratory-derived toxicity figures to field conditions have ignored the excessive amounts probably taken, stressing only minimum lethal doses of chemicals. The illusion of safety thereby created (sometimes, I think, intentionally) very likely conflicts with actual hazards.

Secondary poisoning of mammals, particularly canids, is a usual consequence of field rodent control campaigns. Foxes, coyotes, and domestic dogs are particularly susceptible to indirect 1080 poisoning. Many control officials believe that ground squirrel poisoning with Compound 1080 is the cheapest and most effective approach to coyote control. Shortly after field trials with 1080 began, Kalmbach (unpublished) estimated that rodent control may effect a 30 per cent reduction in the coyote population of treated areas. Wallace, more recently (cited in Longhurst *et al.*, 1952), believed that as much as 90 per cent of a coyote population is removed in some areas by ground squirrel control. Whatever the estimate of reduction, the regular occurrence of secondary poisoning must be accepted as fact. Foxes and skunks have also been found dead in treated areas, and it is presumed that poisoned rodents were responsible. On these and other small furbearing mammals there is no good study that would indicate the extent of population loss. It is, however, probably less severe than that among coyotes, and is more apt to be localized in its expression.

The actual extent of secondary poisoning occurrences is not known. Only in the coyote example do we have a measure of its influence on population numbers. I have no doubt that it is more common than current appraisals indicate. The increasing use of Compound 1080 would seem to ensure widening secondary poisoning in mammals. The difficulty of arriving at proper appraisals in a field investigation is obvious to biologists. Such an appraisal is further hampered by the differing viewpoints regarding mammal control and control methods. The agencies responsible for control are the least likely to provide the appraisal.

20 · FOOD CHAINS AND PESTICIDES: TRANSFERRED CHEMICALS AND DELAYED EXPRESSION

Not only can a chemical demonstrably be passed from the tissues of one living thing to another, as we have just seen, but also it can become increasingly concentrated as it is passed through a food sequence. Ultimately it may produce death. The two-link transferral in which both members die we have described as *secondary poisoning. Delayed expression*, as I use it here, describes the transferral along two or more links of a food chain in which only the terminal member dies. The conditions that allow secondary poisoning also apply here, although, rather uniformly, an initially less toxic chemical is in question. Too high a toxicity does not spare the first link in the chain. In delayed expression, tissue contamination, but not death, must occur in at least the first one or two links. Because the initial toxicity is lower, a chemical must be concentrated at each successive step in its transfer for ultimate toxicity. Each member that concentrates a toxic chemical in its tissues is spoken of as a *biological concentrator*.

To review our definitions for clarity: Secondary poisoning concerns two animals, both of which die but neither of which necessarily concentrates a toxic chemical in its tissues. Delayed toxicity also concerns two members, the first of which may not die and the second of which dies only when the chemicals stored in tissues are mobilized during periods of environmental or physiological stress. Delayed expression normally adds at least one more member to the food chain, and it is only this last one that shows symptoms of poisoning. The poisoning may be lethal to this member or to its young.

Delayed expression as used here emphasizes death of the terminal member. It should be clear that the sequence in most instances is more complex than described here. Biologically, a food relationship is required with the earlier links unresponsive to stored chemicals, and all links showing a marked degree of obligation to particular food types. Sublethal symptoms possibly appear early in the sequence, but they have not been observed.

The delayed expression of toxic symptoms among biological concentrators of pesticides is a serious problem primarily because it is insidious. It goes unnoticed until mortality results, and even then the cause of mortality may often not be suspected. The problem becomes all the more serious when delayed expression is traced back to the chemical-control practices from which it arose: Approved and recommended procedures have produced it. Most applications are at low rates shown to be "safe" or at worst to be minimally hazardous. All employ the stable chlorinated hydrocarbon insecticides in widespread use. In many situations, repeated use of these chemicals is the rule. In some instances, particular programs using stable insecticides extend over millions of acres annually, and, when coupled with the known accretions over many years of use, constitute wide and continuing sources of ecological contamination. The illustrations below were barely hinted at twelve years ago, and only in the past few years has their significance been established. Delayed expression is a result of the general distribution of stable pesticides. I have no doubt that it is more profound and more widespread than the examples below indicate.

The greatest hazard accrues with repeated applications of stable chemicals. Time between applications, dose rates, and the stability of residues must be taken into account in assessing hazard in specific situations. Biologically, concentration is most likely in "closed" ecosystems, that is, where obligate feeding relationships exist in environments that effectively confine chemical residues. Stable bodies of water are the most probable sites, but, as will become clear, residue concentration can occur in other sites where conditions are met.

The economic significance of delayed toxic expression of biologically concentrated chemicals will vary with region, intensity of land use, diversity of fauna, and attitudes. It ranks with insecticide resistance as one of the greatest biological problems yet to appear from chemical pest control. It is particularly important to the wildlife biologist. Unfortunately, wildlife does not share the legal controls that presumably ensure safe limits to human tissue contamination. In view of the general occurrence of contamination with pesticide residues, I must also express doubt that human "safe limits" are properly recognized. Certainly the processes by which both human beings and wildlife acquire tissue contaminants are the same. Only the amounts differ.

New approaches will be required to determine the nature and economics of delayed expression. Specific investigations will necessitate a greater variety of specialists, all of whom must appreciate ecological complexities. More money and time must be devoted to investigations of this kind, since short-term work on isolated facets cannot lead to comprehension of the general problem. A drastic reorientation will be needed in pesticidal field investigations before the full significance of delayed expression can be assessed.

Six examples of delayed expression, three each from aquatic and terrestrial communities, are described below. They are not equally supported by fact, but ecologists will agree, I am sure, that the sequences are reasonable. Only small gaps in information remain to be filled; the evidence for delayed expression of transferred chemicals is no longer tenuous. Heading each example is a simplified diagram indicating the presumed pathways of transferred chemicals. We may be quite sure that chemical effects ramify more widely than is shown, but there is yet no evidence confirming that this is true. The point at which mortality occurs is indicated by boldface capital letters. Direct mortality, where it occurs immediately following chemical application, is also shown in boldface type, but will not be discussed here.

Delayed Expression in Aquatic Communities

Example 1.—The death of fish-eating birds following DDD applications to control gnats at Clear Lake, California.

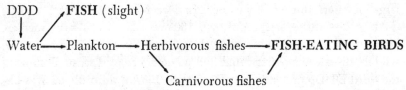

DDD ⟋ FISH (slight)

Water ⟶ Plankton ⟶ Herbivorous fishes ⟶ **FISH-EATING BIRDS**

Carnivorous fishes

Clear Lake is a shallow, warm body of water of about 46,000 acres, 100 miles north of San Francisco. Its high productivity is reflected in excellent sport fishing, one of the attractions that has made Clear Lake an important tourist and recreation center. Unfortunately, insects also share this high productivity. The Clear Lake gnat (*Chaoborus astictopus*) particularly qualifies as a serious nuisance. Although it is not blood-sucking, its numbers during the warmer months are so great that it becomes an intolerable nuisance. Efforts to control the gnat have been made for many years, but it was not until 1948 that an effective method not hazardous to fish was found. The target zone was the bottom muck of the lake in which the gnat larvae develop. DDD —a near relative of DDT—was applied to the surface of the lake and mixed in as much as possible. A concentration of one part DDD in seventy million parts of water was sufficient to kill larvae. It also killed large numbers of other aquatic invertebrates, but did little immediate harm to fish populations. An application of DDD was made to the lake in September of each of the years 1949, 1954, and 1957. The second and third applications were at the higher rate of one part DDD to fifty million parts of water. The first *Chaoborus* kill was 99 per cent, but within three years numbers were again high. In 1954 a second application also produced a 99 per cent kill. Numbers recovered yet more rapidly after this application, and a third, in 1957, was not so successful in *Chaoborus* control. Adverse weather was one reason offered, but the rapid recovery of gnat numbers and the lesser immediate reduction suggest resistance to the insecticide.

Before the first application more than 1000 pairs of western grebes (*Aechmophorous occidentalis*) bred at the lake; next year the summer breeding colony did not return. To the time of this writing it has not been re-established. A few pairs (15–20) of birds were present during the breeding seasons of 1958–60, but no young were brought off. In 1961, 16 nests of western grebes were found, but no young were fledged. One young bird was seen in 1962. Three young were observed in 1963. Grebes had continued, nonetheless, to visit the lake in large numbers during the winter. In December, 1954, death among these grebes was widespread; over 100 were known to have died. A later die-off occurred in March of the following year. In both instances, wildlife-disease biologists could find no evidence of disease. Following the third DDD application, in December, 1957, a third die-off was observed. Disease was also looked for on this occasion, and not found. As an afterthought, the visceral fat of two grebes was removed and analyzed for DDD. The astounding level of 1600 parts per million was found in these tissues—a concentration 80,000 times as great as that applied to the lake. Contaminated food was judged to be the source. This conclusion initiated the collections that were to make clear the sequence of biological concentrators, which perhaps constitutes thus far the best example of the delayed toxic expression of transferred chemicals. Collections of many kinds of fishes and other organisms have been made since 1958, and will continue until no trace of DDD can be found in tissues. Upon recommendations of the Lake County Mosquito Abatement District, no further applications of DDD to the lake have been made.

Hundreds of specimens—plankton, fish, frogs, and birds—from Clear Lake have now been chemically analyzed. All contained DDD. Analyses center on fishes, for it is in them that most chemical concentrating takes place. But birds are not the only fish-eaters; so also are many thousands of the people who visit Clear Lake annually.

In general, the range of DDD in these fishes is 40 to 2500 parts per million. Most DDD is found in the fatty tissues, which represent the main source of contamination. Birds, of course, eat whole, usually small, fish. DDD values recorded from visceral fat (in parts per million) therefore provide a minimum measure of contamination to birds. The ranges of observed contamination are in Table 17.

Human feeders discard a large portion of the fish, including some parts where DDD is most concentrated. We might expect therefore that human fish-eaters would be spared very high levels of contamination, and indeed they are. To assess hazard to human beings, only DDD in edible flesh need be considered. But the levels of DDD in edible flesh are still substantial, and many are above the maximum tolerance level of 7 p.p.m. set by the Food and Drug Administration for DDD residues in marketed foodstuffs. Only commonly eaten species are listed in Table 18. Cooking reduced values by only about 10 per cent.

Analyses of the fatty portion of edible flesh naturally yielded much higher values than did entire edible flesh.

Contamination is not equal in all fishes. Those that are plankton feeders or herbivores store less, as do younger fish of all categories. Storage is maximal and average contamination greater among predaceous fishes. To illustrate, largemouth bass which hatched seven to nine months after the last DDD application showed a range of DDD in edible flesh of 22 to 25 p.p.m.; fish a year older, 25 to 30 p.p.m., and six-year-old fish, an average of 138 p.p.m. Sacramento blackfish, which are plankton feeders, showed values of 7 to 9 p.p.m. in young which hatched several months following the third application of DDD. Old blackfish, in contrast, had two to three times this range in edible flesh. These values indicate only that human beings do ingest residue-contaminated flesh. They give no real measure of the amount of the residue eaten. Fish-eating birds and fishes will, of course, consume larger amounts of DDD.

TABLE 17

Concentration of DDD in visceral fat of Clear Lake fish

Species	Concentration (p.p.m.)
White catfish	80–2375
Largemouth bass	1550–1700
Brown bullhead	342–2500
Blackfish	700–983
Bluegill	125–254
Carp	40
Frogs	5

TABLE 18

Concentration of DDD in edible flesh of Clear Lake fish

Species	Concentration (p.p.m.)
White catfish	1–196 *
Largemouth bass	4–138 †
Brown bullhead	12–80
Bluegill	5–10

* Most above 20.
† About half above 20.

The initial source of contamination in the food chain is plankton. At Clear Lake, floating algae are often so abundant as to color the surface of the water. A composite plankton sample had a contamination value of 5.3 p.p.m.

In the following sequence I have taken some license to arrive at single numbers describing the factorial increase in concentration over the original source of contamination in a simple food chain. These are maximal estimates. We begin with a calculated contamination of water at 0.02 p.p.m. DDD. In vertebrates, only the concentration of DDD in visceral fat is considered.

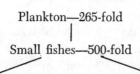

Plankton—265-fold

Small fishes—500-fold

Predaceous fishes—85,000-fold   Predaceous birds (grebes)—80,000-fold

Warm-blooded animals are more susceptible to ingested poisons than are cold-blooded. However, it is only the route of intake—ingestion—that spares the fishes. Gill-breathing animals are particularly susceptible to chlorinated hydrocarbon insecticides in the surrounding water. A fivefold to tenfold increase of DDD in the water (i.e., 0.1–0.2 p.p.m.) would be quite enough to kill all fish in the lake. The relatively low mortality to fish following standard applications of DDD is *prima facie* evidence that the water itself does not contain immediately hazardous amounts of DDD. Although no fish deaths have been observed from these high levels of stored DDD, one may not conclude that no hazard exists. The likelihood of further contamination to threshold toxic levels has not been explored, nor have the effects on breeding success or survival of young.

A great deal more remains to be learned from this example, even though no further contamination from chlorinated hydrocarbon insecticides is planned. For that matter, the effects on fish-eating birds were probably far greater than pictured here. Only western grebes received systematic attention, yet the populations of egrets and herons are also much lower than in former years. Logically, these too must have gone the way of the western grebe, but only indirect evidence brings us to this conclusion. Pesticide-wildlife problems are too often defined as narrowly as the pest control procedures responsible for them.

The Clear Lake accounting, incomplete as it is, still stands as a fascinating example of the complexities of chemical transferral in living systems, and as a pointed warning to professional workers not to oversimplify their solutions to pest control problems. (Data for this section come from Hunt and Bischoff, 1960; Lindquist and Roth, 1950; and unpublished material of Hunt and Bischoff and of A. H. Miller.)

Example 2.—The death of fish-eating animals following applications of toxaphene at 2 pounds per acre to land for insect control, residues subsequently finding their way into water.

Toxaphene (others?)

Land→Water→Algae→Zooplankton→Fishes→**FISH-EATING VERTEBRATES**

An unusual mortality in many species of fish-eating birds was first noticed in May, 1960, on Klamath Basin refuges administered by the U.S. Fish and Wildlife Service. Within a short time 307 birds were found dead. The majority were picked up on the Lower Klamath Refuge on the northern California border. The numbers and species —all fish-eaters—were distributed as shown in Table 19.

TABLE 19

Species distribution of dead birds found on Lower Klamath Refuge

Species	Number
White pelicans	156
American egrets	84
"Seagulls"	34
Black-crowned night herons	12
Western grebes	12
Double-crested cormorants	5
Great blue herons	3
Snowy egrets	1

Wildlife-disease experts were immediately called in for consultation, but no disease symptoms could be found. Alerted by the Clear Lake experience and having observed apparent symptoms of chlorinated hydrocarbon poisoning, the experts had chemical analyses made of tissues of the two species most commonly found dead. The liver of an American egret was found to contain 50.2 p.p.m. of toxaphene. The liver and kidney of a white pelican contained contamination levels of 2.9 p.p.m. and 23.4 p.p.m., respectively. The suspicion became stronger that poisoning, not disease, was the cause of this die-off.

The predominant fishes in ponds and drains at these two refuges are cyprinid minnows, chubs of the genera *Gila* and *Siphateles*. Other genera are present, but the two named seem to be the favored foods of most piscivorous birds. Water level and flow in the refuge are closely regulated, tied to agricultural irrigation on and off the refuges. A considerable amount of refuge waters flow from surrounding cultivated land.

Pesticides had been used in and around the refuges for many years. DDT had been used for 19 years, and aldrin and toxaphene extensively in recent years, though over lesser acreages. Toxaphene had been aerially applied in 1958 at two pounds per acre to control cutworm on a 1500-acre tract bordering the main breeding areas. Other chlorinated hydrocarbon insecticides had been commonly used in adjoining agricultural lands. Therefore, both the source and the occurrence of insecticide poisoning were established.

A series of poisoned birds was subsequently analyzed for tissue contamination. Most contained toxaphene. The kidneys, for example, of a white pelican analyzed contained 14 p.p.m.; an entire American egret contained 17 p.p.m. toxaphene. Grebes, terns, and herons also contained toxaphene in amounts ranging up to 39 p.p.m. But residues of other insecticides were discovered in tissues. The American egret just mentioned had, in addition to toxaphene, 138 p.p.m. DDT and 52 p.p.m. DDD, or a total of 207 p.p.m. of chlorinated hydrocarbon residues. The subcutaneous fat of a western grebe contained 39 p.p.m., 348 p.p.m., and 302 p.p.m. of toxaphene, DDT, and DDD, respectively, a total of 689 p.p.m. Other birds showed similar patterns of contamination. Although separation of effects from the various toxicants was virtually impossible, toxaphene was judged to be the chief cause of mortality because of its higher toxicity and its known application within the refuge in large amounts two years before. Disease was ruled out as a cause because of the absence generally of disease symptoms and the observation of poisoning symptoms (Keith, personal communication).

Thus far we have concluded that deaths resulted from pesticide poisoning, probably from tissue accumulations of toxaphene and possibly with supplemental effects from other contaminants. Can we conclude that delayed expression of transferred chemicals was the probable route of exposure? The first fact to recall is that the nearest heavy application of insecticide occurred almost two years before. From other experience with toxaphene, one may conclude that at least three-quarters of the toxaphene residues had disappeared within a year's time. Even if this fact is discounted, an application of 2 pounds per acre would not produce residue levels in soil or on foliage of the amounts recorded in bird tissues. Biological concentration had to be responsible; confirmation awaited chemical analysis of earlier links in the food chain. In due course, analyses were made. Simplified, they are presented in Table 20.

TABLE 20

Maximum concentrations of pesticides in Klamath Basin organisms, given in parts per million (total weight)

Organism	Amount of residue (p.p.m.)
Algae	0.1–0.3 toxaphene
Snails; *Daphnia*	0.2 toxaphene; 0.7 DDE (DDT by-product)
Small fish	3.0 toxaphene; 0.2 other
Larger fish (6 inches)	8.0 toxaphene
Fish-eating birds	39.0 toxaphene; 138 DDT; others also

These values appear to be much below those recorded in the Clear Lake study. Had visceral fat samples been used throughout, the values would have been more nearly comparable.

The probable sequence was further documented by tissue contamination levels found in animals with different feeding habits:

Plant-feeding birds (coots)	Trace or no contamination
Birds feeding on small aquatic and terrestrial animals (eared grebes)	0.5–4.0 p.p.m. toxaphene
Fish-eating snake (garter snake)	3.2 p.p.m. toxaphene
Fish-eating birds	2–56 p.p.m., including two or more residues

Particularly thought-provoking was multiple contamination of the ovaries of a double-crested cormorant. For the entire bird the residue contamination was 2.2 p.p.m. toxaphene and 2.4 p.p.m. DDE. Developing eggs in the ovary of this bird, however, contained 20 p.p.m. toxaphene and 12 p.p.m. each of DDT and DDE. The bird was apparently healthy when shot. At this time we can only surmise the possible adverse effects of multiple high-level contamination on this bird's breeding abilities. We do not know the extent of similar contamination throughout any wild population, nor could we discriminate with assuredness chemically caused impairments of functions from those caused by other factors. I have no doubt that such residue levels have important, albeit disguised, effects (data chiefly from Pillmore, 1961).

Example 3.—The loss of birds following two toxaphene applications to eliminate fish from Big Bear Lake, California.

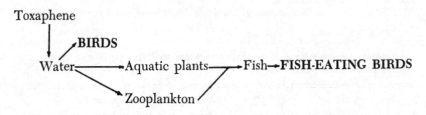

In 1960 toxaphene was applied twice to Big Bear Lake in San Bernardino County, California, in attempts to eliminate populations of "rough fish" (2 species of goldfish) inhibiting trout fishing. The two aerial applications brought the total rate of application to 0.1 p.p.m. in the shallower portion of the lake and 0.2 p.p.m. in the deeper. Toxaphene is a very effective fish poison, and at these concentrations most fish were killed. Shortly after the application a number of birds died. Sick and dead birds were reported for a period of weeks thereafter.

In this instance mortality apparently came from three causes, whose separate contributions are difficult to establish. Direct poisoning apparently caused the death of fish and some birds. Within a few days of the poisoning a noticeable film of oil covered portions of the lake. The oil presumably came from the bodies of decaying fish. The oil film itself apparently water-logged many birds, and it may well have contributed to a greater assimilation of toxaphene. The likelihood of delayed chemical expression is suggested by the high values of toxaphene found in tissues. Over two hundred birds were found dead. Some fifty dead coots probably died from acute poisoning. No dead birds were analyzed chemically. Living birds, shot and analyzed, had tissue contamination (in parts per million) as shown in Table 21.

Arranged in food sequences, toxaphene contamination increased in this fashion (given in p.p.m.):

Aquatic plants (*Potamogeton*)	0.03
Zooplankton (*Daphnia*)	0.73–0.97
Fish flesh	3–21
Birds, fat	13–1700

Of two gulls shot and analyzed, one had been banded in Montana and the other in Idaho. This indicates, as is now recorded in so many instances, that migratory birds are an important means of dispersing chemical residues widely.

TABLE 21

Tissue contamination in surviving birds near Big Bear Lake

Species	Toxaphene in tissue (p.p.m.)
White pelican, fat	1700
White pelican, liver	6
Ducks, fat	12–69
Gulls, fat	39
Eared grebes, fat	33
Terns, fat	13

Trout could not be introduced into the lake for sixteen months following the application; repeated trials showed that the water continued to remain toxic to fish ("hot," as the fisheries biologists put it). Meanwhile, the study of contamination continues. It is probable that delayed expression contributed to observed mortality and may yet cause additional mortality. However, trout introduced into the lake in late 1962 survived; no poisoning of birds was observed in 1963 (data unpublished; from E. G. Hunt).

Delayed Expression in Terrestrial Communities

Example 1.—Mortality among birds many months after the application of DDT at 2–5 pounds per tree to control bark beetles.

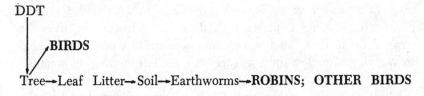

DDT

BIRDS

Tree→Leaf Litter→Soil→Earthworms→ROBINS; OTHER BIRDS

Only a few years ago the accounting I am about to give would have seemed incredible. Now, both the fact of extensive losses of birds and the routes by which loss is caused are carefully documented. The only unknown features remaining are why pest control workers failed for so long to concede that loss was occurring and what the actual losses in bird numbers must be. The truth has finally penetrated control circles, but the total losses of birds cannot be known. Probably they range into millions. Among current control programs only the fire ant control campaign can have had effects as great.

The data now available are too voluminous to record here in detail. A reader should certainly investigate the following references to appreciate the importance of this problem: Barker, 1958; Benton, 1951; Blagbrough, 1952; Hickey and Hunt, 1960; Hunt, 1960; Mehner and Wallace, 1959; and Wallace *et al.*, 1961. Other work in progress will swell the documentation. The main emphasis here will be on the chain of chemical transferral leading to the delayed expression of toxic symptoms and mortality in birds. Other aspects of the problem are treated in only a general way.

The fungus that causes Dutch elm disease was discovered in the United States only 30 years ago. Since the early recognition of the disease in Ohio and New York, it has swept through some 20 states, leaving behind millions of dead and dying elm trees. DDT gave promise of controlling the two species of bark beetles primarily responsible for the spread of the fungus. Dutch elm disease control programs were quickly put into effect by many communities, with some local success. Nonetheless, the affliction continued to spread. In 1949 it had reached Illinois; in 1950, Michigan; in 1956, Wisconsin; in 1957, Iowa. Removal of affected trees controls the disease effectively, as does also the less toxic methoxychlor. Unfortunately, both of these control methods are more expensive than DDT applications. Most organized community programs in the Middle West depend on DDT for control, following state and federal recommendations for formulations, dosages, and methods of application. Foliar sprays are no longer used; dormant sprays are depended upon. From the point of view of chemical transferral it makes little difference whether the trees are sprayed in April or October, but most communities now have turned to spraying after leaf-drop in early fall and, depending on weather, throughout the winter. The conditions for delayed expression remain.

Although bird mortality from DDT spraying of elms had been recorded in 1947, not until 1950 were the observations begun that led to discovery of the transferral of DDT through biological concentrators in a food chain (Barker, 1958). In 1949, when the University of Illinois campus at Urbana-Champaign was first sprayed to control phloem necrosis, very few birds died. Dutch elm disease, however, appeared shortly thereafter and required additional spraying. Several months later dying robins were noticed on the campus before the spring spraying program began. Dying birds were frequently reported after rains; hence, it was suspected that water puddles on contaminated soils contained DDT. Analyses of the water showed only traces of DDT. Food was turned to next as a possible source of DDT. In the summer of 1950, earthworms were collected for analysis, and in due course the presence of DDT in them was chemically established. Many chemical analyses were made in the next two years, and these confirmed the cycle of chemical transferral diagramed at the beginning of this section.

A series of robins and other vertebrates analyzed for DDT and its metabolic derivative DDE established the presence of chemical residues in a number of internal organs. In Table 22 are a few examples of concentrations in organs of animals found dead or dying.

TABLE 22

Concentrations of DDT and DDE in internal organs of vertebrates following bark beetle control

Tissue	Species	Amounts (p.p.m.)	
		DDT	DDE
Brain	Robin	28	11
	Robin	90	252
	Rock dove	4	175
	House sparrow	167	29
	Starling	0	29
	Gray squirrel	4	0
Liver	Robin	19	744
Heart	Robin	127	120

Chemical analyses of various organs in earthworms collected after early summer spraying revealed DDT in all. The range was from 4 p.p.m. in the ventral nerve cord to 403 p.p.m. in the crop and gizzard. All organs except the ventral nerve cord contained levels greater than 48 p.p.m. The analyses of various annelid worm species showed comparable levels of contamination in all. Six specimens collected before the first spray in 1951 (8 to 9 months after the nearest prior spraying) had average overall residue contaminations of 86 p.p.m. of DDT and 33 p.p.m. of DDE.

The theory that DDT was the chemical agent causing death and that the earthworm was the toxic carrier seemed nearly proved. A chemical cycle needed only its intermediate steps confirmed. Ultimately these were also established. A complete sequence of contamination, transferral, concentration, and mortality could now be constructed. The steps may be arranged in this order:

Elm tree	2 lbs. or more of DDT per tree in summer
Soil surface	1.3 lbs. per acre
Top two inches of soil	5.9–9.5 p.p.m. DDT; 0.7–5.4 p.p.m. DDE
Leaves following spraying	174–273 p.p.m. DDT; 9–20 p.p.m. DDE
Leaves after leaf-drop, several months later	20–28 p.p.m. DDT; 1–4 p.p.m. DDE
Earthworms, immediately after spraying	33–164 p.p.m. DDT; 3–73 p.p.m. DDE
Earthworms, several months later	86 p.p.m. DDT; 33 p.p.m. DDE

Robins returning for nesting in the spring following the summer's sprayings were therefore presented with a favored food that had concentrated DDT by at least a factor of 10 over that present in soil. In other terms, a diet of 100 earthworms contained more than 3 mg. of DDT, a lethal dose to a robin. The individual organs of the robin possibly concentrated the chemical slightly more, but for general comparisons the levels of contamination in earthworm and robin were the same. Presumably the more complex nervous system of the robin succumbs more swiftly than that of the earthworm to insecticides whose primary action is on the central nervous system. Yet, the population of earthworms seems to be declining on the University of Illinois campus. Possibly they too have their limits.

The facts of mortality and of delayed chemical expression at its root being established, the next question is extent of population loss. A mortality offset by good nesting success will produce no declines in average yearly numbers. But an excessive mortality combined with no or poor nesting success can result only in rapid decline in population size.

Precipitous declines among many species of birds have in fact been the rule where Dutch elm disease control with DDT has been repetitively practiced. Wallace *et al.* (1961) record a decline in numbers of robins on the main campus of Michigan State University, dropping in three years from 370 to 15 to 4 birds. Nesting failures were more startling; virtually no broods were brought off during the years after spraying began. Hickey and Hunt (1960) estimate total robin mortality to be at least 85 per cent for each spraying. In several Wisconsin communities consecutively sprayed for three years, total breeding-bird populations were reduced by 30 to 90 per cent. Where there were as few as 10 elm trees per acre and these were sprayed annually, the drop in population was about 90 per cent.

Although robins seem to be the most obvious victims of DDT treatments for Dutch elm disease control, many other species are affected and there is no doubt of widespread declines in affected avifaunas. Wallace *et al.* (1961), for example, have compiled a list of 94 species of birds found dead or dying in treated areas. Samples of 41 of these species have been analyzed; 34 of the species contained DDT. Foremost among the affected species are the ground feeders, followed closely by foliage-feeding orioles, warblers, and vireos, and less commonly by bud-feeders and bark-foraging species. Curiously, several species of carnivorous birds were also found dying with tremors, and chemical deposits were found in many individuals. Screech owls were commonly found dead or dying, but no one can say precisely what the route of poisoning was. Secondary poisoning is a possibility, but it seems more likely to me that this species at certain times of year also feeds on earthworms, and therefore its route of poisoning would be similar to that of robins. This conclusion is circumstantially supported by the observation that dead screech owls were found after a period of unusually heavy rains which brought large numbers of earthworms to the surface (Hickey, personal communication). The death of large hawks and owls cannot yet be understood. But, whatever the feeding group or the manner of poisoning, it can be convincingly demonstrated that bird populations have been widely reduced.

The question, as frequently posed, is not simply one of choice between the elm trees or the birds. Advocates of the cheapest existing control methods present it in this way, but they make a basic error in doing so. Neither safeguarding trees nor safeguarding birds can be justified on strictly economic grounds, but both are valued for the comfort and pleasure they afford; both must be safeguarded. If necessary, more expensive measures must be called upon to do this. The search for acceptable, economical, and simple control methods must go on for the best interests of all. Unquestionably, DDT treatment has resulted in saving many elms and in slowing the spread of Dutch elm disease. But the treatment has also resulted in rapid declines in bird populations. Control purposes should never be defined so narrowly as to hazard general advantage, whatever sanctity can be assumed within the limits of single purpose.

Example 2.—The reduction of reproductive success in woodcock following repeated sprayings of forested areas in New Brunswick at 0.5 and 1 lb. of DDT per acre.

DDT
|
Forest and marsh——►Earthworms(?)——►WOODCOCK (YOUNG)

Presumptive evidence suggests that fewer young woodcock are being brought off in DDT-sprayed areas of New Brunswick than in un-sprayed areas (Wright, 1960). Treatment to contain spruce-budworm outbreaks has covered many millions of acres in the last decade. At the application rate of 0.5 lb. per acre aerially applied, there seem to be no effects adjudged seriously intolerable by anyone and there is no question that insect numbers are curbed at least annually. The serious aspects of the problem are the effects on both woodcock and budworm of repeated sprayings. Permanent suppression of budworm numbers does not follow single annual treatment, and woodcock pop-ulations, able to recover the following year from a single annual dis-ruption of food sources, seem not to be able to sustain numbers over a succession of treatment years. Some 22,000 square miles have been sprayed in New Brunswick since 1952, and most areas have been sprayed at least twice. Some areas have received as many as five treatments, although these have seldom been annual retreatments.

Wright (1960, and personal communication), taking the only method of analysis available to him, has compared the numbers of adult and immature birds in the fall bag of woodcock from sprayed and un-sprayed areas. An indirect measure of this kind admittedly cannot establish pesticide effects on reproductive suppression. However, the observed differences are statistically significant and the pattern of de-cline of young emerging from the annual series of data seems more than merely interesting.

In the years 1953 to 1958, the number of immature birds comprised roughly half the total number of birds shot in unsprayed areas. In

1953 the figure for the sprayed area was the same as that for the unsprayed area, but the spraying took place after young birds had hatched. In the next three years the numbers of immature birds taken from sprayed areas were markedly reduced. Expressed as percentages, they composed successively 17, 21, and 20 per cent of fall bag counts, in contrast to comparable bags of 50, 52, and 45 per cent of immatures from unsprayed areas. Over the four-year period, sprayed areas produced 23 per cent immature birds in bag, whereas 45 per cent immatures were counted in bags from unsprayed areas. The production of young on sprayed areas was, by this measure, one-half what it was on unsprayed areas. The most convincing evidence that this reduction is meaningfully associated with DDT comes from the final year, 1959, in which no spraying was conducted. During this year immature birds in bag from previously sprayed areas increased to 49 per cent of the total, an increase of 145 per cent over the prior year. Spectacular as is this rate of increase, it still lagged behind the percentage of immature birds taken from unsprayed areas in that year. These constituted roughly 60 per cent of the fall shoot. Evidently total recovery is not possible even after one year's freedom from DDT exposure.

A pattern of delayed chemical expression, in this instance producing indirect effects on young, has not been convincingly established among these woodcock populations. I offer the example here because I believe that research now under way will bear it out. The biology of the species and the pattern of insecticide application make it a reasonable supposition.

A complicating factor has recently been introduced. In 1959 a June sample of 10 woodcock showed an average of 0.18 p.p.m. of heptachlor epoxide in tissue. Some two-thirds of the North American population of woodcock winter in the southern United States, much of which was treated with heptachlor in the fire ant control campaign. Earthworms taken from this area as much as 12 months after treatment contained an average of 3 p.p.m. of heptachlor in tissue. Only 10 per cent of woodcock which had recently arrived at their treated wintering grounds contained detectable traces of heptachlor or its epoxide. But six weeks after arrival more than 50 per cent contained an average of 0.2 p.p.m. (DeWitt *et al.*, 1960). The woodcock is now being subjected to pesticide contamination at both ends of its 2000-mile migratory pathway. The two chemical residues in foods or tissues, acting in consonance, may yield more drastic effects than yet seen. Only time and concerted attention will tell this. Meanwhile, double chemical jeopardy to an important game species with rather narrow feeding habits remains a serious possibility.

Example 3.—Mortality and population declines among vertebrates in the southern states on areas treated with heptachlor (primarily) at 1.5 lbs. per acre in efforts to eradicate the imported fire ant.

Heptachlor

```
                        VERTEBRATES
Marsh and field ──→ Earthworms(?) ──→ WOODCOCK(?)

          Unknown ──────→ GROUND-DWELLING VERTEBRATES
```

The fire ant control program has resulted in heavy immediate mortality to vertebrates, declines in population numbers, slow rates of recovery, and residual contamination of tissues with heptachlor and heptachlor epoxide in most of the fauna in the treated area. These facts are established; yet it is difficult to establish that delayed expression occurs in contaminated populations. In spite of much research investigation, the many efforts of the U.S. Fish and Wildlife Service and cooperating agencies have lacked an ecological core. The focus has been on population counting and tissue analyses, both of which aid in explaining how and to what extent biological processes are impaired by the intrusion of toxic chemicals but neither of which should be considered a terminal goal of investigation. Nonetheless, all the conditions for delayed chemical expression are present under this control regime; I fully expect that many examples of chemical transferrals along food chains will become clear in time.

Earthworms on woodcock wintering grounds in Louisiana contained up to 20 p.p.m. of heptachlor epoxide in tissue 6 to 10 months after land treatment. As just noted, woodcock reflect these concentrations by storing significant amounts of the epoxide in their own tissues. As yet, there is no indication of direct mortality from this source of contamination.

Carnivorous mammals contained residues that must have come largely from contaminated foods. Cotton rats, for example, which serve as an important food for carnivores, had a residue level of 5.9 p.p.m. heptachlor epoxide in tissue. Fishes and amphibians, which are fed upon regularly by vertebrates, almost without exception had detectable amounts of heptachlor epoxide. Reptiles too were commonly contaminated. In all these groups that serve as food, the source of tissue contamination must be their own food species. A centrarchid fish (bluegill), for example, with a residue level in tissue of 36.2 p.p.m. could have obtained such concentrations and still survived only if the epoxide was gradually accumulated from food sources. Crayfish with 1.6 p.p.m. in tissue after months of exposure could have acquired it only through ingestion of lesser amounts. One-tenth of that concentration in water would effectively obliterate entire populations of both bluegill and crayfish. Food-chain transferral of chemicals must be one of the commonest sources of effects following survival to the initial exposure. The final linkage of transferred chemicals and long-delayed effects has yet to be made, however (data from DeWitt and George, 1960).

21 · FAUNAL DISPLACEMENT—
SPECIES OUT OF BALANCE

The term faunal displacement denotes a change in the numbers and distribution of single species, or a rearrangement of the component species in an ecosystem. The number and character of single populations normally vary from season to season and year to year. Sometimes the changes are great, but more often, all other factors being equal, changes throughout the year and over the years are reasonably predictable and set by the limitations of environment. Changes in abundance, distribution, and faunal composition come about naturally, in either long or short term, through the many factors previously discussed. No population increases indefinitely, and the rule in undisturbed environments is a relative stability in faunal systems whose members have evolved together and have adapted, each to the other's presence.

The "balance of nature" concept has arisen from this commonly observed stability. But no one suggests that such a balance is absolute or that a once-disturbed environment will inevitably revert to the precise composition and stability of the former one. What we have often taken to be the natural stabilized population is only the population of our experience. As a rule this population in man-occupied regions is the latest product of a sequence of disturbances and returns to stability. Our reasons for wishing that a disturbed area be returned to its former condition may have little to do with the question, "Does a balance of nature exist?"

In broad terms, it does, of course, exist. Confirmation of its existence derives most certainly from the consequences of the biotic simplification we induce on croplands and from the results of translocations of species from one region to another, whether accidental or intentional. Imbalance and instability follow either, and species translocations compound the effects of biotic simplification. Whenever additional factors further simplify already unstable environments, the balance is further upset. Living with this upset is one of the major trials of man. Pest control is one response of man to try to rectify an imbalance of nature.

However, pesticides are the most recent general contributors to faunal imbalance. Attempts to stabilize with chemicals have in fact guaranteed instability. The present chapter centers on pesticidal contributions to faunal instability. Nonchemical pest control measures can also yield instability, but not of such magnitude. Any group of animals within a faunal complex responds when major shifts in representation occur in other groups. But only in arthropods are the consequences of such shifts at all understood. Pest insects naturally have formed the major core of studies on these shifts.

In the majority of cases, pesticides used on croplands do not cause imbalances that cannot be contained for the short term. Pesticide applications that do cause economically or biologically serious imbalances may have these consequences in use:

(1) Many species are reduced in numbers along with the pest; the natural enemies of the pest are frequently killed along with the pest species.

(2) Repeated application may produce resistance to the chemical.

(3) Pesticides may be ineffective against potentially damaging species but instead may remove controlling species, allowing the former to increase to pest proportions.

(4) Nonselective chemical applications ensure a need for more frequent treatment, with catastrophe threatened if treatments are withheld; a decreasing dependence on natural controls is the rule.

(5) Costs increase and a greater likelihood of inacceptable residues and occupational hazard results from ever-increasing dependence on chemicals.

(6) Among vertebrates, selective removal of predators may allow prey populations to increase to pest levels.

For convenience, faunal displacement can be divided into two time scales. Short-term effects are chiefly those that occur during a given crop year, demanding for their alleviation the attention of many applied biologists. The most important faunal problem created by pesticide applications is the resurgence of numbers of pest species. Recovery of numbers is frequently rapid after treatment, and may often lead to populations larger and more destructive than those before treatment. Long-term effects are those that become obvious only after a period of years, creating hazard not easily assessed at any given time. Their gradual rise to serious status is, too frequently, marked by failure to recognize that any change is occurring at all. It is the insidious manner as well as the ultimately obvious severity that distinguishes the longer-range effects of pesticidal applications (general reference, Ripper, 1956).

Resurgences in Numbers Following Chemical Treatment

In spite of high initial kill, pesticides at times bring about a tremendous increase in numbers of the species against which they were applied. Many of the nonselective chlorinated hydrocarbons and organophosphates do this, but the effect is not limited to these chemical groups. Less toxic kinds, such as sulfur fungicides and the "inert" diluents of insecticidal formulations, can also produce a resurgence— or "flare-back," as it is commonly called. Resurgences have been recorded in both temperate and tropical climates for over 50 species of plant-eating insects and mites. The majority of these species are serious pests now found widely in comparable crop types throughout the world.

Spraying in orchards has repeatedly intensified pest numbers. Pickett and co-workers, in a series of papers (e.g., Pickett *et al.*, 1958), have described this phenomenon in apple orchards in Nova Scotia. They report, for example, that extensive use of lime-nicotine sulfate dust between 1918 and 1924 led to extreme outbreaks of the eye-spotted budmoth and to the appearance of red spider mite. The codling moth increased in numbers following use of calcium arsenate. Oyster-shell scale became a major problem following use of mild sulfur fungicide. Red spider outbreaks followed DDT applications. These workers list many more examples, all showing that established pests often increase and new pests may appear after pesticide applications.

The same phenomena have been observed in citrus culture. Malayan citrus orchards regularly treated with kerosene emulsion had greater populations of citrus blackfly than did untreated orchards; scale insects were more numerous in Florida orchards treated with sulfur; mite and aphid populations increased noticeably after DDT use in California orchards; citrus red mite increased so drastically in Florida after DDT applications that all recommendations for DDT use in citrus orchards were withdrawn (DeBach and Bartlett, 1951).

In field crops as in orchards, repeated use of many common insecticides and fungicides has led to outbreaks of mites, aphids, and other arthropods. Attempts to control the devastating spotted alfalfa aphid with parathion permitted more severe outbreaks (Stern and van den Bosch, 1959). The cyclamen mite on strawberries increased dramatically following treatment with parathion (Huffaker and Kennett, 1956). Treatment with nonselective insecticides of aphids on vegetable crops resulted in resurgences requiring retreatment within two weeks. When para-oxon, for example, was used commercially in England, there occurred within 14 days the greatest cabbage aphid outbreak ever recorded, even though initial mortality had been extremely high (Ripper, 1956). The use of heptachlor in the fire ant control program in Louisiana led to both increased populations of and damage by the sugar cane borer (Hensley *et al.*, 1961). Additional sequelae of this program were several lawsuits.

In forests, displacements of this kind do not seem severe. Hoffmann and Merkel (1948) noted temporary reductions in numbers, but no permanent displacements of either target or nontarget species. In forests we may conclude that resurgences are made less likely by the generally lighter dose rates, the relative infrequency of application, and the innate complexity of the faunal system. Nonetheless, in New Guinea forest following a cessation of repeated aerial treatments with DDT to control a malarial mosquito, numbers of this vector rose sharply (Laird, 1959). A key to the difference in response is apparently repetition.

In forest soils, resurgences seem not to occur (Hartenstein, 1960), but in agricultural soils higher populations of collembolans are the rule following treatment with DDT, BHC, and parathion (Satchell, 1955). Again, repetition seems to be the critical difference.

In waters, too, resurgence occurs among aquatic insects. When a nuisance species is the first to return in numbers following aerial spraying of forests with DDT, it masks the true situation and the conclusion is often drawn that no permanent reductions in the numbers of all insects occurred. Rather, it is a change in balance, with the nuisance species comprising a greater percentage of the total population than before. Davies (1950), for example, showed that the application of DDT to an Ontario stream to control simuliid blackflies led to a dramatic and continuing increase in their numbers; three years after application the average emergence rate was 17 times as great as in the precontrol period. Following large-scale treatment of forests, black-flies are regularly the first to recover and most likely to become numerically dominant (see Chapter 6). Total removal of competitive forms, not repetition of treatments, seems to underlie this response.

In all environments, resurgences in numbers can follow repeated pesticide treatments. Usually associated with these resurgences are broad-spectrum insecticides. It now remains to examine what biological factors permit resurgences in pest populations.

There are several possible explanations. All involve in some way an alteration of the biotic potential, rendering rapid return to high numbers more likely. The resurgence may be through a directly favorable influence of the pesticide on the individual, but more often the influences are indirect.

The state of nutrition may be influenced, but nutritional studies are often insufficient to allow correlations with population phenomena. I mentioned earlier that small amounts of some pesticides are nutritionally valuable. We can think of these as "animal fertilizers." Possibly, too, the enhanced welfare of the plant is reflected in increased vigor of the survivors of insecticide treatment. The reduction in insect numbers may provide the survivors with less competition for more and choicer foods. Laird (1958) extended this idea to pesticide effects on larval mosquitoes in culture media. He showed that DDT in the medium was responsible for an initial high mortality of larvae. Dead larvae hastened bacterial action, which in turn hastened the maturation of the microfauna. The consequently enriched culture satisfied the food requirements of growing larval mosquitoes much more adequately than did normal cultures not treated with DDT. The product was an insect more robust than normal. But this was conducted in the laboratory, and therefore, as with most studies of nutritive contributions to animal welfare, field testing is needed.

Resistant strains are perhaps better considered a long-term consequence of insecticide exposure (see Chapter 11). Here I wish to comment only that resurgence in numbers is an inevitable result of insecticide resistance. The question has arisen whether resistant strains are biologically more vigorous than unresistant strains. One result, if this were true, would be that resistant insects would increase in numbers beyond those of pretreatment populations. There is some argument about this point. In a few instances resistant strains of insects seem to fare better—lay more eggs, have shorter generation times, and so on. Yet there are many instances in which resistant strains performed no more ably, indeed many performed more poorly, than their unaltered counterparts. We shall leave the point open, but suggest strongly that any unusual resurgence among a resistant population is more apt to be due to a change in the faunal balance than it is to an inherent biological advantage accruing from insecticide resistance.

The lessening of competition, related indirectly to nutrition, is better demonstrated as a factor in increasing the welfare of both individuals and populations. All living things suffer the effects of crowding. A reduction in numbers frequently leads to compensatory responses in growth and reproductive success. Simple reduction of numbers, therefore, will in most groups lead to more rapid population recovery.

But all of these factors are of less consequence in permitting resurgence than is the reduction of natural enemies. In arthropods, perhaps less so in vertebrates, this is the factor that explains most observed instances of population recovery following chemical treatment.

One of the most dramatic examples implicating pesticides as destroyers of the natural enemies of pests occurred with the advent of DDT use in citrus orchards in 1946 (DeBach and Bartlett, 1951). The singular success of the vedalia beetle, imported from Australia to control cottony cushion scale, is one of the classic tales of biological control. For almost 60 years this scale had been kept in check, at low numbers, by its introduced adversary, but with widespread applications of DDT the scale increased sufficiently to become a serious pest once more. DDT was much more toxic to the coccinellid predator than it was to the scale. Only the uniform withdrawal of DDT from citrus orchards restored the former balance.

Instances are now legion in which DDT and parathion unfavorably depressed predator and parasite populations. It is inferred that this suppression of natural enemies, combined with their inability to reproduce as quickly as their prey, permits the rapid resurgence of pests. Inference may not be enough to establish the case. Two general methods of appraisal may be used to verify that resurgence in fact depends on pesticidal inhibition of parasite and predator numbers.

The first method is to compare the composition of sprayed and unsprayed plots with regard to frequency of preying insects. There are many reports of this kind. A few examples follow. Spider populations were higher in unsprayed New Jersey apple orchards than in sprayed. The aggressive, or hunting, spiders were more seriously affected than the passive, or web-building, species (Specht and Dondale, 1960). Native predators play a significant role in keeping aphid populations in check in California alfalfa fields. Malathion, as one broad-spectrum insecticide, practically eliminates the important coccinellid predator *Hippodamia* and the aphid-feeding syrphid flies. Repeatedly, aphid numbers increased rapidly where these predators had been reduced to low levels by this nonselective chemical (Smith and Hagen, 1959). Predaceous mites were commoner in unsprayed Connecticut apple orchards than in sprayed (Hurlbutt, 1958). Doutt (1948) recorded that DDT applications in pear orchards were definitely responsible for severe infestations of the Baker mealybug, through suppression of natural enemies, and that, moreover, DDT residues remained lethal to these enemies for long periods, essentially serving as a barrier to predator recovery for an entire crop season.

A variation of this comparative method is to correlate change in frequency of pest injury to level of predator population. One example in which this variant method was used is the sugar cane borer increase (Hensley *et al.*, 1961), previously cited in this chapter. Pickett (1961) has employed the same method by noting reduction of apple scab in orchard-insect complexes freed of repetitive chemical intrusions. In 1948, for example, before the beginning of an integrated program employing both chemical and ecological methods, 73.1 per cent of fruit escaped injury; in 1959, with the concept and practice of integrated control fully installed, 93.9 per cent escaped insect injury.

The second method of checking on the interrelatedness of pest populations and their natural enemies is to exploit resurgence as an experimental method. Since some arthropods increase following the use of insecticides, what more effective way of establishing the significance of predation than by deliberately eliminating it as a controlling force! DeBach (1946) first described the "insecticidal check" method, and it has been used many times since. Fleschner (1958), for example, showed that citrus red mite populations were much higher on DDT-treated trees than on citrus from which predators of the mite had been removed by hand. And mite populations were, of course, much lower where there was no removal of predators by either means.

Several experimental studies having important theoretical as well as practical conclusions have appeared since the insecticidal check method came into being. Perhaps the best of these is that of Huffaker and Kennett (1956). These workers used laboratory populations of a plant-feeding mite and its predator, also a mite, and followed with field trials to confirm their results. Their three basic methods were removal of predators with parathion, removal of predators by hand, and stocking of predators. Each approach confirmed the others. The plant-feeding mites were twenty times more numerous where predators were absent than where predators were stocked. The resurgence of the pest mite to excessive numbers was easily producible by the parathion application.

Changes in Faunal Composition

Long-range changes in arthropod complexes follow from the continued use of nonspecific chemicals, whatever their chemical nature. One might argue that long-range effects do not differ in kind from effects occurring within single crop years. But that is not entirely true. The analysis of faunal shifts during single crop years does not permit prediction of long-range directional changes in fauna, while a long-range analysis can facilitate predictions of directions of change in other situations. Rather than to say that short-term and long-term effects differ in kind, it may be more proper to say that they differ in prognosticative value: We do not have the knowledge to predict future directions from short-term studies. Unfortunately, applied even to the short term, faunal analyses at best describe few species and few of the factors that influence change in these species. The unstudied portions may be the key to predictability. Though multivariate analyses in nature merit much more time and attention, the standard limitations of money, personnel, and purpose dictate that there will be no great change in the predictability of long-range faunal shifts. Therefore, we must piece together patterns from fractional evidence, i.e., depend largely on a historical approach, to confirm that repetitive chemical use over the years does in fact alter faunal composition. The discussion that follows necessarily centers on economically important consequences of faunal change. The diagram below illustrates the kinds and directions of change I have been discussing.

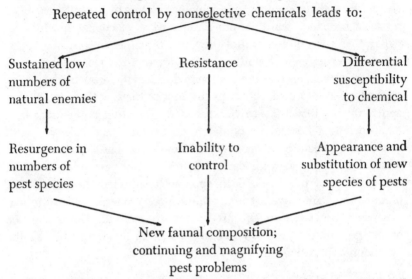

Repeated control by nonselective chemicals leads to:

Sustained low numbers of natural enemies

Resistance

Differential susceptibility to chemical

Resurgence in numbers of pest species

Inability to control

Appearance and substitution of new species of pests

New faunal composition; continuing and magnifying pest problems

The cumulative result of chemical treatment applied over a period of years is a fauna poorer in parasites and predators of types able to cope with strong resurgences of pest numbers. The natural enemies that remain fluctuate greatly in numbers. The sequel to the resulting imbalance is chronic reappearance of pest outbreaks. Outbreaks occur more frequently, often during poor climatic conditions, and a new pest condition, man-induced, is the product. The normal grower reaction is to apply chemicals more frequently. Following more frequent exposure, the beginnings of resistance appear in the pests, manifested by decreased ability of the chemical to kill at usual application rates. Then not only is the frequency of application raised, but also the dose. If, too, certain species are less susceptible to usual chemicals and increase their numbers to damaging levels, these species will require treatment with different chemicals. Often two or more are combined in single applications. The result of all three grower reactions is to magnify outbreaks still further. Chemical treatments reach their limits of utility when full-fledged insecticide resistance sets in, when the cost of increasingly frequent applications becomes prohibitive, or when, no longer able to sustain loss, the grower switches to other crops or means of control.

Resurgence is at the root of the grower's problems; resistance complicates control; and differential susceptibility, coupled with destruction of natural enemies and competitors, raises to pest status a species formerly rated minor or not considered a pest at all. The first two aspects of the sequence have already been discussed. The last also requires discussion, for it is often important. The new pests generated by continued chemical application are sometimes called substituted, or secondary, pests.

Shortly after the advent of DDT, mites became serious pests in many orchard crops because of their relatively low susceptibility. Other chemicals, some less toxic, have also given rise to secondary pests. For example, sulfur sprays contributed to the increasing pest status of eye-spotted budmoth, oyster-shell scale, and European red mite in Nova Scotian orchards (Pickett *et al.*, 1958). Soft brown scale, formerly a minor pest, increased drastically after parathion applications (DeBach and Bartlett, 1951). Many more examples could be cited, but all show that a species once not considered a pest became one. The destruction and continued suppression of natural enemies by nonselective chemicals explain the new pest status. Several years may be required before a pest or group of pests becomes serious, and as Pickett and his colleagues have shown, efforts to lessen damage from a pest complex through integrated chemical and ecological control may also require years.

The tetranychid mites provide some of the best illustrations of the formation and development of substituted pests (secondary only in time, not damage). The diagram following pictures the evolution from insignificance to major pest of one important mite species (*Metatetranychus ulmi*) in England as a result of a succession of pesticide exposures of different nature and purposes (Jacob, 1958). Needless to say, the sequence is generalized.

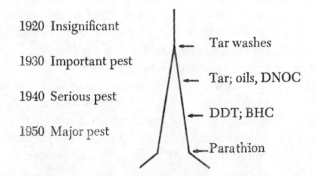

1920 Insignificant	
	← Tar washes
1930 Important pest	
	← Tar; oils, DNOC
1940 Serious pest	
	← DDT; BHC
1950 Major pest	
	←Parathion

The bulk of pest-arthropod studies do not give the kinds of information needed to assess substitution or replacement phenomena in faunal complexes. A focus on the single species adjudged serious at a given time precludes study of entire complexes. It is now becoming clear that inordinate efforts against single species with pesticides that affect many are conducive to continuing pest problems. The awareness of this possibility has centered in three major entomological nuclei: western European foresters and orchardists; Canadian orchard ecologists (Pickett and his group primarily); and workers in biological control, notably Canadian forest entomologists and the group of entomologists at the University of California. In 1939, A. J. Nicholson had formalized the concern that pesticides might be producing pests. Postwar synthetic chemicals have convincingly demonstrated the reality of this concern.

Recently, replacement phenomena among insects of public-health importance have received attention from thoughtful students (Aitken and Trapido, 1961; Laird, 1959). The studies of Aitken and Trapido on Sardinia will serve to illustrate.

Eradication attempts against the important malarial vector *Anopheles labranchiae* led to new dominance relationships within the entire anopheline mosquito assemblage. Shifting numbers of five species were studied over a five-year period, beginning with the eradication program against *labranchiae* in 1946. After 1948, *labranchiae* became uncommon, but another species, *A. hispaniola,* previously rare, became common. Intensive checks over wide areas in 1952 confirmed that *hispaniola* had effectively taken over the niche vacated by eradication attempts against *labranchiae.* Evidently, *hispaniola* was able to continue its dominant position, once established. Fortunately, *hispaniola* is not a malarial vector. We see here, therefore, an important possibility—the substitution of an innocuous species for a hazardous one by planned ecological management. Yet *A. sacharovi,* a very effective malarial vector, was commonly found in 1952 while seemingly not being present before that year, and three other species of anophelines became commoner during the years of eradication treatments. Clearly these anopheline species are in dynamic balance with one another, and even a presumably selective attack on a single element in the complex unwittingly affects the others. Disease transmission may be either mitigated or intensified by these realignments, but whatever the assessment of pest hazard, we should be alert to changes in faunal constitution whenever pesticides are directed against single species.

Discussion of this nature can be applied to vertebrate as well as arthropod populations, but the data in faunal terms are more difficult to assess. Pesticide effects on vertebrates concern fewer species, fewer numbers, longer generations, and more complex behavior patterns. Moreover, vertebrate animals are rarely approached dispassionately. In view of this greater interest in and the relative ease of studying vertebrates in contrast to arthropods, I have thought it best to describe them separately. Thus, in prior sections on food, competition, and predation, I discussed resurgences in populations and successional shifts of species compositions among vertebrates. Normally, the hub of such discussions was a single species considered to be economically important. Nonetheless, I should like to see the new emphasis on faunistic management by integrated control measures, now becoming prevalent among entomologists, paralleled among the manipulators of vertebrate populations. Vertebrate pest control and wildlife management suffer from entirely too much preoccupation with single species.

VI · RETROSPECT AND PROSPECT

22 · POINTS OF VIEW AND THE FUTURE

Throughout this book I have held to the premise that the intentions and consequences of chemical pest control as currently carried out are defined inadequately. My purpose has been to describe the full scope of pest control by identifying, first, the land-use practices that favor the increase and spread of pest species; second, the means by which pests are combated; third, the biological community into which these controls are introduced; fourth, some short-term and long-term effects of controls on the fauna. I have specifically emphasized how continued heavy dependence on present chemical-control systems must ultimately be self-defeating; how control methods have departed from fundamental biological principles; and finally, how the limited goals of present pest control practices fail to serve either longer-range productive values or the varied interests of a public which is called upon to support these practices and which is inextricably drawn into the consequences.

Whatever the immediate consequences of toxic chemicals in the "living landscape"—and they are sometimes serious—it must be said that they are normally correctable. The simplest counter to a chemical practice that produces immediately obvious undesirable effects is to end the practice before its effects become irreversible. But the less obvious consequences—and many are profoundly important—are less easily handled and are not quickly recognized for what they are. The biological specialists from whom recommendations for chemical use come must soon examine more profoundly the long-range consequences of these recommendations.

Let me review what I consider the general problem to be before I go on to some specific proposals for change. Crop systems are necessarily ecologically simplified for economical production of marketable foods and fibers. The frequent consequence is an unbalanced ecosystem in which a few species of organisms increase beyond thresholds of numbers that we can tolerate. By definition, these species then become pests. Where present knowledge does not provide cultural or biological practices that counter the effects of "created" pests, we must use chemicals to reduce numbers temporarily. Our purpose in using chemicals in this fashion should be only to "buy" time enough to work out effective and permanent solutions to pest problems. Supposedly, chemicals are so selected and applied as to accomplish only this purpose without further complicating problems of pest control or producing toxic effects beyond the target species. These are logical expectations of pesticide uses in agriculture, where most pesticides are applied.

In forestry, range management, and public health, the requirements for pest control do not directly parallel those in agriculture. In these disciplines dependence on environmental control is much greater. Chemical controls are generally instituted only when it appears that habitat manipulation will fail. The reasons for such failure are many, and are often social rather than biological. Although the gains from many chemical-control practices in these areas may reasonably be expected to be economic, less tangible purposes are served as frequently. However subjective the values entering the decision to control with chemicals, the suppositions I described above for agricultural use apply here also. Specificity of chemical action is a logical expectation wherever chemicals are used. In those instances where specificity is not precise and unwanted damage results, the character and amount of this damage must be known in advance and carefully weighed against presumed gains from chemical application. The decision to initiate control should therefore be taken with the fullest possible knowledge of its purposes and consequences as defined by the entire spectrum of impinging biological, economic, and social values. The action of a toxic chemical must be specific and temporary; but the value to man of the chemical in use must be general and lasting.

Present chemical pest control practices are often an incongruous mixture of actual need, superficial appraisal, unbalanced perspective, and expediency. Complaints against these practices in use are lodged against, first, the chemicals themselves—their hazards in use; then the pest control practitioners—their failure to appreciate the biological complexities of their work; and, finally, the ends served—the lack of full consideration for all values influenced by chemical use.

The first of these complaints is, in one sense, the least serious. The fact that there is direct hazard from toxic chemicals is acknowledged by a variety of testing, educational, regulatory, and legislative agencies. There is an awareness in government, industry, and agriculture that pesticidal chemicals are potentially harmful. The question remains whether this awareness is great enough and whether it is focused sufficiently on hazards to species other than man. Broad-spectrum chemicals in use always kill more than the pest species; yet this fact is lightly regarded unless complaints are made. Too often, a given recommendation for the practical use of a pesticide centers on a simple one-to-one relationship: This chemical can kill that pest. An illusion is created that the effects of the chemical are indeed confined to its target. That is by no means true, either of the direct killing propensity of a toxic chemical or of the indirect ecological effects that must follow its use. Chemicals that are sufficiently stable to leave toxic residues for extended periods are participants in a further illusion: that toxic residues are actually "controlled" by the recommending and regulatory procedures now in force. Present residue legislation neither protects consumers from contamination nor sufficiently recognizes that residues do not behave as we would wish them to. Chemicals are not precisely applied; they are not confined only to areas needing treatment. Residues do accumulate in soils and living tissue. Through both physical and biological channels they do "escape" from the areas where the parent chemicals are applied. It is my view that the scope of hazard from toxic chemicals is generally defined too narrowly. An illusion of safety follows.

There is a heartening stirring in the ranks of applied biologists. The steady drift of the last two decades away from recognition of the fundamentally biological nature of pest control is now being countered by some strong new voices. Problems of hazard, resistance, and residue control have given substance to the belief that generalized chemical pest control is in the long run self-defeating. Simmons (1957) advanced this thought with regard to disease vectors, but his conclusions may be generally applied. He put it this way:

It is evident that the extermination of vectors with residual insecticides is probably not feasible. We appear to be in an endless cycle of synthesizing, at an ever increasing cost, more and more insecticides to which vectors become more and more resistant. Thus, it seems we cannot go on forever relying on insecticides. A more final and permanent solution to the vector-borne disease problem might be in ecologic control.

Geier and Clark (1961) came to similar conclusions for crop-insect control with chemicals and suggested that many current programs are little more than ". . . a series of blows delivered almost blindly but with a maximum of violence." This they described as the "fly-swat approach," keeping most trained personnel occupied in applied work with little time and energy for thinking about fundamentals.

Pest control continues to lack fundamental character. Its methods are chiefly empirical and its goals are too narrowly circumscribed. For too long a time the apparent success of chemical controls seemingly obviated any need for full comprehension of biological relationships in pest control.

Misplaced emphasis on chemicals also modifies assessment of the nature and values of advantages brought about by pest control. The prospectuses and reports of chemical-control programs conducted by public agencies swell with pronouncements of general advantage. Coincident with the decline of dependence on biologically based pest control has been a decline in the consideration of hazards and side-effects inherent in chemicals in use. The devaluation of side-effects and impinging interests has coincided with a period of ever-expanding chemical control programs. In other terms, though chemicals produce more side-effects over more area, these are weighed less by control personnel than in former years. The net benefits of present programs of reduction, containment, and eradication are not convincingly demonstrated, nor is any serious effort made to justify such programs on economic grounds. Furthermore, matters of residues, wildlife loss, biotic imbalances, and long-term effects are often accorded short shrift as merely nuisance considerations. Alternative methods of control are usually met with apathy or are dismissed as ineffective or impractical. The price of chemical dependence in pest control has unfortunately been the dilution of values once considered important and the acceptance of limited purposes as standard goals.

In the immediate future, I expect to see increasing public resistance to programs depending on indiscriminate application of toxic chemicals. Contamination of the "living landscape" will continue to mount and will be found to be serious at sites not now suspected. New routes of harm will be discovered, and those now only suspected will be established as serious. Increasing evidence will support a resistant public, which, if only intuitively, demands of scientists and public servants more sophisticated solutions to complex problems and wider responsibility for their actions than now characterize pest control.

A number of courses of action have been suggested in specific contexts in earlier sections. More generally defined, I would urge these immediate actions (also see Appendix, with whose recommendations I concur):

Employ the "eradication" philosophy only on recent and spatially limited invasions of new pests, or only where the techniques of eradication are clearly selective.

Stop all chemical treatments that depend on high doses of stable chemicals whose effects are not confined to the target species. At present this conclusion applies most cogently to programs employing dieldrin, aldrin, and heptachlor. By biological definition these programs are control failures, and I suspect that they are economic failures as well. The only exception to this recommendation would apply when catastrophe is imminent and obvious; no control program in the United States now fits this category.

Encourage cultural and faunal diversification programs that ensure more complex faunas and less stability of pest populations. Immediately required are accelerated efforts in biological control, the selection of resistant plants, and diversification of the landscape.

Initiate a "user education" program in which the full range of hazards of chemical use is explained. It is not enough to seek to educate growers and appliers in the proper methods of handling toxic chemicals. They must also be told the biological and social consequences of continuing chemical use. Alternatives to chemical use in both the short term and the long term must be as clearly explained as immediate hazard to the applier now is.

Require public justification of any tax-supported control program and insist that this justification contain a clear description of the purposes of the control program, the methods by which it is to be conducted, the hazards inherent in it, and the economic or social gains that justify both costs and hazards.

There are two additional areas of action, which will require longer periods but cannot fail to yield more generally satisfactory pest control. The first of these is a changed outlook on the part of pest control practitioners; the second is the modification of existing pest control administration toward a broader responsibility in pest control.

Traditional pest control must yield to a new outlook, one that does not treat pest species as somehow independent of the natural forces influencing all living things. Viewed in this light, the conservation of one species and the protection against undue numbers of another are two sides of the same coin. The management of population numbers, for whatever purpose, requires close knowledge of the ecological factors that determine population density. We must therefore adopt the view that pest control is no more than the conscious lowering of numbers of undesirable species and that the word "pest" itself has no meaning when populations are small enough to cause man no important inconvenience. Biological manipulations, as I have shown, are much surer means to the end of achieving control of pests than are chemical applications.

Good beginnings have already been made. They are reflected in the terms "supervised pest control" and "integrated" or "harmonious" pest control. Geier and Clark (1961) call for "protective population management." I have elsewhere urged "ecological management," and, moreover, have advised that the ecological manager be considered the primary source of pest control recommendations. Pest control should be viewed only as a special application of population ecology. There is abundant evidence that the regulation of numbers by ecological manipulation is reasonable and sure. Although not all current pest problems can be quickly solved by these methods, the promise is great. It is economically expedient now to replace traditional pest control with population management, but gains in the long run go well beyond economics.

Adoption by applied biologists of this new outlook on control will persuade pest control agencies and organizations to administer policies consistent with ecological facts. Traditions die slowly, however. Several actions might be followed to hasten the transition to saner control programs and more judicious use of chemicals.

First among these is more active coordination among governmental agencies bearing responsibility for the regulation of pesticides or for their use in publicly conducted control programs. Intended to serve a coordinating function is the Federal Pest Control Review Board, instituted in 1961. Within the framework of traditional pest control methods and customary alignments of responsibilities within particular agencies, this board, I am sure, represents an improvement over previously existing coordinating instruments. It is nonetheless only a first step; it will provide a greater efficiency in the conduct of control operations by federal agencies, but will change little the underlying methods and philosophies. These too require change. Accordingly, the purposes of this board should be redefined to encompass the necessary and desirable changes I have previously described. It is not enough simply to increase the efficiency of operations whose methods should be discarded. To assist in formulating new definitions, representation on this board should be widened to include all interests affected by pest control practices. This will require representation by industry, educational institutions, and conservation organizations. The federal agencies by no means adequately represent all interests in pest control.

Training, coordination, and integration of pest control groups normally are oriented toward producing greater efficiency in control programs and wider representation of interests but do not necessarily take into account that changes in attitude are required. It is my belief that existing knowledge of the hazards of toxic chemicals in nature is quite enough to warrant fundamental changes in outlook. This view is shared by many biologists and resource specialists.

Doubts about the justification for major change can be dispelled by judicious appraisals from universally acceptable bodies. The present Committee on Pest Control and Wildlife of the National Research Council must, I believe, conclude that wildlife hazards are more serious than we have previously acknowledged. By its mandate, however, it cannot go far enough. I would suggest that the National Research Council's Committee on Biology and Agriculture and Committee on Natural Resources jointly assume responsibility for a general appraisal of the significance of environmental contamination by toxic chemicals. Their appraisal could begin in the context of a yet larger study—that of man-induced changes in producing ecosystems. Prominent among the themes of such a study must assuredly be the biological consequences of ever-widening monocultural planting with its resulting faunal simplification. An attempt must be made to understand both process of change and change itself in producing ecosystems on a national scale. Only then will we be able to understand the full importance of environmental contamination with pesticides, and only then will we be able to protect fully the health of man himself. Even now, without complete understanding, the manifold effects of pesticides constitute a major issue of our time and one becoming more serious with each passing year. Understanding is obscured and responsibility diluted by the present practice within the National Research Council of fragmenting resource problems (subcommittees reporting to subcommittees).

I have treated this subject essentially as an American issue, but pest control, considered as a problem in the management of living resources for the best interests of man, knows no national boundaries. The environment is contaminated with toxic chemicals wherever pest control is conducted throughout the world and the concomitant problems are remarkably similar, irrespective of differences in the environments and the cultures of native peoples. The magnitude of these problems is becoming clear in a few countries in which toxic chemicals are used extensively. With expanding chemical use, we can fully expect problems to become more serious throughout most producing areas of the earth.

International agencies have thus far treated chemical uses only in the context of immediate goals and limited definitions of purpose. The World Health Organization, for example, has been preoccupied with the chemical control of disease vectors, and concern for side-effects is largely limited to immediate human hazard. The Rockefeller Foundation, though interested in the same goals, has extended its valuations somewhat further, into the question of what social consequences result from successful disease control. The Food and Agricultural Organization presents pesticides to its members chiefly as instruments of food production. Although immediate hazards of pesticides to human health are well defined by the Food and Agricultural Organization, problems under its charge are not approached from a broad biological base, nor are the widely ramifying effects of pesticides in producing ecosystems acknowledged. Until recently, the International Union for Conservation of Nature and Natural Resources concerned itself primarily with the protection of wild species in game preserves rather than with the manifold problems of ecologically based resource management.

Production and protection are parts of the same cloth. They cannot be biologically separated. I suggest that international agencies begin the study of pest control in a broader context, and that, as a first step, groups be organized to appraise the present impact of chemical contamination of biological resources, whatever the ultimate use of these resources might be. The Rockefeller Foundation, the organization most experienced in problems of international biology and human welfare, might best initiate the appraisals, and include recommendations for appropriate research programs. Other organizations could then assume the responsibility for study, appraisal, and recommendation appropriate to their purposes.

Immediate action is necessary. We can no longer afford to dismiss piecemeal the "separate" problems that arise from uncontrolled chemicals in living environments. There are no separate problems.

**APPENDIX
REFERENCES
INDEX**

APPENDIX

*Recommendations of the President's Scientific
Advisory Committee on Use of Pesticides
(Kennedy, 1963: pp. 19–23)*

*Verbatim transcription of part IV of the President's report on pesticides is
included here primarily because its recommendations deal specifically with
the activities of federal agencies. Although I have discussed many of the
same recommendations in the text, I have usually done so in a different context. The entire report is commended for reading; it is, in my view, the most
judicious appraisal of pesticidal values and hazards yet to appear.*

RECOMMENDATIONS

The Panel's recommendations are directed to an assessment of the levels of
pesticides in man and his environment; to measures which will augment the
safety of present practices; to needed research and the development of safer
and more specific methods of pest control; to suggested amendments or public
laws governing the use of pesticides; and to public education.

   A. In order to determine current pesticide levels and their trends in man
and his environment, it is recommended that the Department of Health, Education, and Welfare:

1. Develop a comprehensive data-gathering program so that the levels of pesticides can be determined in occupational workers, in individuals known to have been repeatedly exposed, and in a sample of the general population. As a minimum, the survey should include determinations on fat, brain, liver, and reproductive organs in adults and infants; examinations to determine if placental transmission occurs; and determination of levels which may be excreted in human milk. These studies should use samples sufficiently large and properly drawn to obtain a clear understanding of the manner in which these chemicals are absorbed and distributed in the human body.

2. Cooperate with other departments to develop a continuing network to monitor residue levels in air, water, soil, man, wildlife, and fish. The total diet studies on chlorinated hydrocarbons initiated by the Food and Drug Administration should be expanded. These should, for example, include data on organophosphates, herbicides, and the carbamates in populated areas where they are widely used.

3. Provide Federal funds to assist individual States to improve their capabilities for monitoring pesticide levels in foods which are produced and consumed within the state.

B. In order to augment the safety of present practices, it is recommended that:

1. The Food and Drug Administration proceed as rapidly as possible with its current review of residue tolerances, and the experimental studies on which they are based. When this review is completed, it is recommended that the Secretary of Health, Education, and Welfare select a panel from nominations by the National Academy of Sciences to revaluate toxicological data on presently used pesticides to determine which, if any, current residue tolerances should be altered. Of the commonly used chemicals attention should be directed first to heptachlor, methoxychlor, dieldrin, aldrin, chlordane, lindane, and parathion because their tolerances were originally based upon data which are in particular need of review. Upholding the same standards, the Secretary should ensure that new compounds proposed for registration be rigorously evaluated.

2. The existing Federal advisory and coordinating mechanisms be critically assessed and revised as necessary to provide clear assignments of responsibility for control of pesticide use. The Panel feels that the present mechanisms are inadequate and that it is necessary to provide on a continuing basis for:—

(*a*) Review of present and proposed Federal control and eradication programs to determine if, after consideration of benefits and risks, some programs should be modified or terminated.

(*b*) Development and coordination of a monitoring program conducted by Federal agencies to obtain timely, systematic data on pesticide residues in the environment.

(*c*) Coordination of the research programs of those Federal agencies concerned with pesticides.

(*d*) Initiation of a broad educational program delineating the hazards of both recommended use and of the misuse of pesticides.

(*e*) Review of pesticide uses and, after hazard evaluation, restriction or disapproval for use on a basis of "reasonable doubt" of safety.

(*f*) A forum for appeal by interested parties.

3. The National Academy of Sciences-National Research Council be requested to study the technical issues involved in the concepts of "zero tolerance" and "no residue" with the purpose of suggesting legislative changes.

4. The Secretaries of Agriculture, Interior, and Health, Education, and Welfare review and define their roles in the registration of pesticides that are not present on food, but that may impinge on fish and wildlife or come into intimate contact with the public.

5. The accretion of residues in the environment be controlled by orderly reduction in the use of persistent pesticides.

As a first step, the various agencies of the Federal Government might restrict wide-scale use of persistent insecticides except for necessary control of disease vectors. The Federal agencies should exert their leadership to induce the States to take similar actions.

Elimination of the use of persistent toxic pesticides should be the goal.

C. Research needs:

1. In order to develop safer, more specific controls of pests, it is recommended that Government-sponsored programs continue to shift their emphasis from research on broad spectrum chemicals to provide more support for research on—

(*a*) Selectively toxic chemicals.
(*b*) Nonpersistent chemicals.
(*c*) Selective methods of application.
(*d*) Nonchemical control methods such as the use of attractants and the prevention of reproduction.

In the past few years, the Department of Agriculture has shifted its programs toward these specific controls. The Panel believes this trend should be continued and strengthened. Production of safer, more specific, and less persistent pesticide chemicals is not an unreasonable goal, but its attainment will require extending research efforts beyond empirical approaches to more fundamental studies of subjects such as: the mode of action of pesticides; comparative toxicology; the metabolism of compounds in insects, plants, and higher animals; and the processes of chemical degradation and inactivation in nature. Such studies will also provide the information necessary to control those pests which are rapidly becoming resistant to currently available chemicals. Intensified effort is needed in the search for selective methods of pesticide application. Compounds are often applied in excessive quantity or frequency because of such inefficiencies as drift, uneven coverage, or distribution methods insufficiently specific to reach the target pest.

2. Toxicity studies related to man

The toxicity data upon which registrations and tolerances are based should be more complete and of higher quality. Although data are available on acute toxic effects in man, chronic effects are more readily demonstrated in animals because their generation time is shorter, and thus the natural history of pesticide effects is telescoped chronologically. However, there will continue to be uncertainties in the extrapolation from experimental animals to man, and in the prediction of the nature and frequency of effects in humans on the basis of those observed in other forms of life.

The panel recommends that toxicity studies include determination of—

(*a*) Effects on reproduction through at least two generations in at least two species of warmblooded animals. Observations should include effects on fertility, size and weight of litter, fetal mortality, teratogenicity, growth and development of sucklings and weanlings.

(*b*) Chronic effects on organs of both immature and adult animals, with particular emphasis on tumorigenicity and other effects common to the class of compounds of which the test substance is a member.

(*c*) Possible synergism and potentiation of effects of commonly used pesticides with such commonly used drugs as sedatives, tranquilizers, analgesics, antihypertensive agents, and steroid hormones, which are administered over prolonged periods.

3. Toxicity studies related to wildlife

The Panel recommends expanded research and evaluation by the Department of the Interior of the toxic effects of pesticides on wild vertebrates and invertebrates.

The study of wildlife presents a unique opportunity to discover the effects on the food chain of which each animal is a part, and to determine possible pathways through which accumulated and, in some cases, magnified pesticide residues can find their way directly or indirectly to wildlife and to man.

4. Amplification of research resources

Only by stimulating training and basic investigation in the fields of toxicology and ecology are research needs likely to be met. An increased output of basic research data and a continuing supply of capable research personnel could be ensured by a system of grants and contracts. Training grants, basic research grants, and contracts to universities and other nongovernmental research agencies funded by the Departments of Agriculture, Interior, and Health, Education, and Welfare would stimulate this research. In order to accelerate immediate progress, it might prove useful to explore the contributions which can be made by competent research people and their facilities in other countries.

D. In order to strengthen public laws on pesticides, it is recommended that amendments to public laws be requested. These should:

1. Eliminate "protest" registrations.

The Panel concurs with the Department of Agriculture that these technically evade the intent of the public laws. Industry needs an appeal mechanism, however, to protect it from arbitrary decisions. Public hearings could be held on such appeals.

2. Require that every pesticide formulation carry its official registration number on the label.

The Department of Agriculture has recommended such an amendment as a means of increasing the protection of the consumer.

3. Clarify the intent of the Federal Insectide, Fungicide, and Rodenticide Act to protect fish and wildlife by including them as useful vertebrates and invertebrates.

4. Provide, as a part of the operating budgets of Federal control and eradication programs, funds to evaluate the efficiency of the programs and their effects on nontarget organisms in the environment. Results of these studies should be published promptly.

Approximately $20 million were allocated to pest control programs in 1962, but no funds were provided for concurrent field studies of effects on the environment. The Department of Agriculture has repeatedly suggested that other interested agencies participate in the control programs, but funds have not been available except by diversion from other essential agency functions.

E. To enhance public awareness of pesticide benefits and hazards, it is recommended that the appropriate Federal departments and agencies initiate programs of public education describing the use and the toxic nature of pesticides. Public literature and the experiences of Panel members indicate that, until the publication of "Silent Spring" by Rachel Carson, people were generally unaware of the toxicity of pesticides. The Government should present this information to the public in a way that will make it aware of the dangers while recognizing the value of pesticides.

REFERENCES

ADAMS, J. B. 1960. Effects of spraying 2,4-D amine on coccinellid larvae. Can. J. Zool., **38**: 285–288.

AGRICULTURAL RESEARCH SERVICE. 1954. Losses in agriculture. A preliminary appraisal for review. U.S. Dept. Agr., ARS-20-1, 190 p.

————. 1958. Facts about the imported fire ant eradication program. U.S. Dept. Agr., Unnumbered circ., 10 p.

AHMED, M. K. 1955. Comparative effect of Systox and schradan on some predators of aphids in Egypt. J. Econ. Entomol., **48**(5): 530–532.

AHMED, M. K., L. D. NEWSOM, R. B. EMERSON, and J. S. ROUSSEL. 1954. The effect of Systox on some common predators of the cotton aphid. J. Econ. Entomol., **47**(3): 445–449.

AITKEN, T. H. G., AND H. TRAPIDO. 1961. Replacement phenomenon observed amongst Sardinian anopheline mosquitoes following eradication measures. Proc. 8th Tech. Meeting, Intern. Union for Conserv. of Nature and Natural Resources, Warsaw (1960), p. 106–114. E. J. Brill, Leiden.

ALBERT, A. 1960. Selective toxicity. Methuen, London. 233 p.

ALLEN, D. 1954. Our wildlife legacy. Funk and Wagnalls, New York. 422 p.

ALLEN, N., R. L. WALKER, AND L. C. FIFE. 1954. Persistence of BHC, DDT and toxaphene in soil and the tolerances of certain crops to their residues. U.S. Dept. Agr., Tech. Bull. 1090, 19 p.

300

ANDERSON, B. G. 1960. The toxicity of organic insecticides to *Daphnia*, p. 94–95. *In* Biological problems in water pollution. Trans. 2nd Seminar Biol. Problems in Water Pollution (1959).

ANDERSON, L. D., AND E. L. ATKINS, JR. 1958. Effects of pesticides on bees. Calif. Agr., 12(12): 3–4.

ARANT, F. S., K. L. HAYS, AND D. W. SPEAKE. 1958. Facts about the imported fire ant. Highlights Agr. Research, Ala. Polytech. Inst., Auburn, 5(4), 2 p.

ARNOLD, L. W. 1954. The golden eagle and its economic status. U.S. Fish Wildl. Serv., Circ. 27, 35 p.

ARRINGTON, L. G. 1956. World survey of pest control products. U.S. Dept. Commerce. 213 p.

AUDY, J. R. 1956. Ecological aspects of introduced pests and disease. *In* Symposium on the hazards of imported disease. Med. J. Malaya, 11(1): 20–32.

BAKER, M. F. 1963. New fire ant bait. Highlights Agr. Research, Auburn Univ. Agr. Expt. Sta., 10(2), 1 p. in reprint.

BALCH, R. E., F. E. WEBB, AND J. J. FETTES. 1956. The use of aircraft in forest insect control. Forestry Abstrs., 16(4); 17(1, 2), 31 p.

BARKER, R. J. 1958. Notes on some ecological effects of DDT sprayed on elms. J. Wildl. Mgmt., 22(3): 269–274.

BARNES, J. M. 1958. Control of health hazards associated with the use of pesticides. Advances Pest Control Research, 1: 1–38.

BARTHEL, W. F., R. T. MURPHY, W. G. MITCHELL, AND C. CORLEY. 1960. The fate of heptachlor in the soil following granular application to the surface. J. Agr. Food Chem., 8(6): 445–447.

BATES, M. 1956. Man as an agent in the spread of organisms, p. 788–804. *In* W. L. Thomas (ed.), Man's role in changing the face of the earth. Univ. Chicago Press, Chicago.

BENTON, A. H. 1951. Effects on wildlife of DDT used for control of Dutch elm disease. J. Wildl. Mgmt., 15(1): 20–27.

BISHOP, E. L. 1949. Effects of DDT mosquito larviciding on wildlife. III. The effects on the plankton population of routine larviciding with DDT. Public Health Repts., 62(2): 1263–1268.

BLAGBROUGH, H. P. 1952. Reducing wildlife hazards in Dutch elm disease control. J. Forestry, 50(6): 468–469.

BOLLEN, W. B., H. E. MORRISON, AND H. H. CROWELL. 1954. Effect of field treatments of insecticides on numbers of bacteria, *Streptomyces*, and molds in the soil. J. Econ. Entomol., 47(2): 302–306.

BOND, R. M. 1939. Coyote food habits on the Lava Beds National Monument. J. Wildl. Mgmt., 3(3): 180–198.

———. 1945. Range rodents and plant succession. Trans. 10th N. Amer. Wildl. Conf., p. 229–234.

BORG, K. 1958. Effects of dressed seed on game birds. VII Nordiska Veterinärmötet, Sekt. B, Rapp. 17, 6 p. in reprint; text in Swedish.

———. 1962. Predation on roe deer in Sweden. J. Wildl. Mgmt., **26**(2): 133–136.

BOSTICK, V. 1956. Is DDT converting some national forests into biological deserts? Nature Mag., **49**(4): 202–204, 219.

BOSWELL, V. R. 1952. Residues, soils, and plants, p. 284–297. *In* A. Stefferud (ed.), Insects—The yearbook of agriculture. U.S. Govt. Printing Office, Washington.

BOYD, C. E., S. B. VINSON, AND D. E. FERGUSON. 1962. A case of insecticide cross-resistance in a vertebrate. Unpubl.

BOYD, C. E., S. B. VINSON, AND D. E. FERGUSON. 1963. Possible DDT resistance in two species of frogs. Copeia, **63**(2): 426–429.

BRIDGES, W. R. 1961. Disappearance of endrin from fish and other materials of a pond environment. Trans. Amer. Fish. Soc., **90**(3): 332–334.

BRIEJER, C. J. 1958. The growing resistance of insects to insecticides. Atlantic Naturalist, July–Sept.: 149–155. Reprinted from Mededelingen Directeur van de Tuinbouw, **19**: 861–866 (1956).

BROWN, A. W. A. 1951. Insect control by chemicals. John Wiley & Sons, New York. 817 p.

———. 1958. The spread of insecticide resistance in pest species. Advances Pest Control Research, **2**: 351–414.

———. 1961. Ecological consequences of the development of resistance. Proc. 8th Tech. Meeting, Intern. Union for Conserv. of Nature and Natural Resources, Warsaw (1960), p. 33–37. E. J. Brill, Leiden.

BROWN, W. L., JR. 1961. Mass insect control programs: Four case histories. Psyche, **68**(2, 3): 75–111.

BUECHNER, H. K. 1950. Life history, ecology, and range use of the pronghorn antelope in Trans-Pecos Texas. Amer. Midland Naturalist, **43**(2): 257–354.

BUMP, G., R. W. DARROW, F. C. EDMINSTER, AND W. F. CRISSEY. 1947. The ruffed grouse. Life history, propagation, management. N.Y. Conserv. Dept., Albany. 915 p.

BUREAU OF CHEMISTRY. 1959 and earlier. Ann. Repts. to Calif. Dept. Agr.

BUTLER, P. A. 1962. Effects on commercial fisheries, p. 20–26. *In* Effects of pesticides on fish and wildlife: A review of investigations during 1960. U.S. Fish Wildl. Serv., Circ. 143.

CALIFORNIA DEPARTMENT OF PUBLIC HEALTH. 1961. Occupational disease in California attributed to pesticides and agricultural chemicals, 1959. Calif. State Dept. Public Health, Berkeley, 30 p.

CAMP, A. F. 1956. Modern quarantine problems. Ann. Rev. Entomol., **1**: 367–378.

CARSON, R. 1962. Silent spring. Houghton-Mifflin Co., Boston. 368 p.

CARTER, M. G. 1958. Reclaiming Texas brushland range. J. Range Mgmt., **11**(1): 1–5.

CHRISTIAN, J. J. 1959. The roles of endocrine and behavioral factors in the growth of mammalian populations, p. 71–97. *In* A. Gorbman (ed.), Comparative endocrinology. John Wiley & Sons.

CHRISTIAN, J. J., AND D. E. DAVIS. 1957. The biological basis of rodent control. U.S. Naval Med. Research Inst., Lect. and Rev. Ser. 57–3: 463–472.

CLARK, L. R. 1947. An ecological study of the Australian plague locust *Chortoicetes terminifera* Waek. in the Bogan-Macquarie outbreak area in N.S.W. Council Sci. Ind. Research Australia, Bull. No. 226: 133–134, 546–547, 552, 558.

CLAWSON, M., R. B. HELD, AND C. H. STODDARD. 1960. Land for the future. Resources For the Future, Inc., The Johns Hopkins Press, Baltimore. 570 p.

CLIFFORD, P. A. 1957. Pesticide residues in fluid market milk. Public Health Repts., 72(8): 729–734.

COMMISSION FOR ECONOMIC DEVELOPMENT. 1956. Economic policy for American agriculture, 40 p.

COMMITTEE ON INSECTICIDES. 1960. Insecticide resistance and vector control. 19th Rept. Expert Comm. on Insecticides. World Health Org., Tech. Rept. Ser. 191, 98 p.

COMMODITY STABILIZATION SERVICE. 1960. The pesticide situation for 1958–1959. U.S. Dept. Agr., 24 p. (Issued annually.)

CONSERVATION FOUNDATION. 1960. Ecology and chemical pesticides. Notes and discussions, Nov. 16–17, 1959. The Conservation Foundation, New York, 35 p.

COPE, O. B. 1960. The retention of D.D.T. by trout and whitefish, p. 72–75. *In* Biological problems in water pollution. Trans. 2nd Seminar Biol. Problems in Water Pollution (1959).

————. 1961. Effects of DDT spraying for spruce budworm on fish in the Yellowstone River System. Trans. Amer. Fish. Soc., 90(3): 239–251.

COPE, O. B., AND P. F. SPRINGER. 1958. Mass control of insects: The effects on fish and wildlife. Bull. Entomol. Soc. Amer., 4(2): 52–56.

COTTAM, C., AND F. M. UHLER. 1940. Birds as a factor in controlling insect depredations. U.S. Fish Wildl. Serv., Wildl. Leaflet (B.S.) 162, 6 p.

COUCH, L. K. 1946. Effects of DDT on wildlife in a Mississippi River bottom woodland. Trans. 11th N. Amer. Wildl. Conf., p. 323–329.

CRAFTS, A. S. 1961. The chemistry and mode of action of herbicides. Interscience Publ., New York. 269 p.

CRAIGHEAD, F. C. 1951. A biological and economic evaluation of coyote predation. N.Y. Zool. Soc. and The Conservation Foundation, New York, 23 p.

CRAIGHEAD, J. J. 1959. Predation by hawks, owls, and gulls, p. 35–42. *In* The Oregon meadow mouse irruption of 1957–1958. Fed. Coop. Extension Serv., Oregon State Coll.

CRISLER, L. 1956. Observations of wolves hunting caribou. J. Mammal., 37(3): 337–346.

CROUTER, R. A., AND E. H. VERNON. 1959. Effects of black-headed budworm control on salmon and trout in British Columbia. Can. Fish Cult., No. 24: 23–40.

DAMBACH, C. A., AND D. L. LEEDY. 1948. Ohio studies with repellent materials with notes on damage to corn by pheasants and other wildlife. J. Wildl. Mgmt., 12(4): 392–398.

DARLING, F. F. 1955. West Highland survey. Oxford Univ. Press, London. 438 p.

————. 1956. Man's ecological dominance through domesticated animals on wild lands, p. 778–787. *In* W. L. Thomas (ed.), Man's role in changing the face of the earth. Univ. Chicago Press, Chicago.

DASMANN, R. F. 1959. Environmental conservation. John Wiley & Sons, New York. 307 p.

DAVIES, D. M. 1950. A study of the black fly population of a stream in Algonquin Park, Ontario. Trans. Roy. Can. Inst., 28(59): 121–159.

DAVIS, D. E. 1957. The use of food as a buffer in a predator-prey system. J. Mammal., 38(4): 466–472.

DEBACH, P. 1946. An insecticidal check method for measuring the efficacy of entomophagous insects. J. Econ. Entomol., 39(6): 695–697.

DEBACH, P., AND B. BARTLETT. 1951. Effects of insecticides on biological control of insect pests of citrus. J. Econ. Entomol., 44(3): 372–383.

DECKER, G. C. 1954. Insects in the economic future of man. Agr. Chem., 9(2): 36–39, 111–117.

DELANEY, J. J. (chairman). 1951 and 1952. Hearings before the House Select Committee to investigate the use of chemicals in foods and cosmetics. Parts 1–4. House of Representatives, Eighty-second Congress. 1867 p.

DETHIER, V. G. 1956. Repellents. Ann. Rev. Entomol., 1: 181–202.

DEWITT, J. B. 1956. Chronic toxicity to quail and pheasants of some chlorinated insecticides. J. Agr. Food Chem., 4(10): 863–866.

DEWITT, J. B., AND J. L. BUCKLEY. 1962. Studies on pesticide-eagle relationships. Audubon Field Notes, 16(6): 541.

DEWITT, J. B., AND J. L. GEORGE. 1960. Pesticide-wildlife review, 1959. Bur. Sports Fish. Wildl., U.S. Fish Wildl. Serv., Circ. 84 rev., 36 p.

DEWITT, J. B., C. M. MENZIE, V. A. ADOMAITIS, AND W. L. REICHEL. 1960. Pesticidal residues in animal tissues. Trans. 25th N. Amer. Wildl. Conf., p. 277–285.

DOUTT, R. L. 1948. Effect of codling moth sprays on natural control of the Baker mealy bug. J. Econ. Entomol., 41(1): 116–117.

DOWDEN, P. B., H. A. JAYNES, AND V. M. CAROLIN. 1953. The role of birds in a spruce budworm outbreak in Maine. J. Econ. Entomol., 46(2): 307–312.

EADIE, W. R. 1954. Animal control in field, farm, and forest. Macmillan, New York. 257 p.

EGLER, F. E. 1958. Science, industry, and the abuse of rights of way. Science, 127(3298): 573–580.

————. n.d. Herbicides. 60 questions and answers concerning roadside and right of way vegetation management. Litchfield Hills Audubon Soc., 23 p.

ELTON, C. 1942. Voles, mice, and lemmings. Oxford Univ. Press, London. 496 p.

————. 1958. The ecology of invasions by animals and plants. Methuen, London. 181 p.

ELVEHJEM, C. A. (chairman). 1960. Report on food and feed additives and pesticides. State of Wisconsin Governor's Special Committee on Chemicals and Health Hazards, Madison. 41 p. (mimeo).

EMLEN, J. T., JR., AND B. GLADING. 1945. Increasing valley quail in California. Univ. Calif., Agr. Expt. Sta., Bull. 695, 56 p.

ENTOMOLOGICAL RESEARCH DIVISION. 1959. Residues in fatty tissues, brain, and milk of cattle from insecticides applied for grasshopper control on rangeland. J. Econ. Entomol., 52(6): 1206–1210.

ERRINGTON, P. L. 1943. An analysis of mink predation upon muskrats in north-central United States. Iowa Agr. Expt. Sta., Research Bull. No. 320: 793–924.

————. 1946. Predation and vertebrate populations. Quart. Rev. Biol., 21(2): 144–177; 21(3): 221–245.

FASHINGBAUER, B. A. 1957. The effects of aerial spraying with DDT on wood frogs. Flicker, 29(4): 160.

FAY, R. W. 1959. Changing aspects of insecticide resistance, p. 134–145. *In* Symposium on Pesticides, Brazzaville, Nov. 9–13, 1959. World Health Org.

FEDERAL AVIATION AGENCY. 1960. FFA statistical handbook of aviation, 1960 ed. 137 p.

FICHTER, E. 1953. Control of jack rabbits and prairie dogs on range lands. J. Range Mgmt., 6(1): 16–24.

FISCHER, C. D. 1956. Pesticides. Past, present, and prospects. Originally appeared in Chem. Week, 79: 17, 18, 20. Collected and distributed by Union Carbide and Carbon Corp.

FISHERIES RESEARCH BOARD. 1963. Annual report, 1962–63. Fish. Research Board Can., p. 29–30.

FITCH, H. S., AND J. R. BENTLEY. 1949. Use of California annual-plant forage by range rodents. Ecology, 30(3): 306–321.

FLEMING, W. E., AND W. W. MAINES. 1953. Persistence of DDT in soils of the area infested by the Japanese beetle. J. Econ. Entomol., 46(3): 445–449.

FLESCHNER, C. A. 1952. Host-plant resistance as a factor influencing population density of citrus red mites on orchard trees. J. Econ. Entomol., 45(4): 687–695.

————. 1958. Field approach to population studies of tetranychid mites on citrus and avocado in California. Proc. X Intern. Congr. Entomol. (1956), 2: 669–674.

FLETCHER, J. 1891. President's inaugural address. Insect Life, 4(1): 11–13.

FOOD AND DRUG ADMINISTRATION. 1958. Protecting crops and consumers. U.S. Dept. Health, Educ., Welfare, Leaflet No. 6, 11 p.

FORMOSOV, A. N., AND K. S. KODACHOVA. 1961. Les rongeurs vivant en colonies dans la Steppe Eurasienne et leur influence sur les sols et la végétation. Proc. 8th Tech. Meeting, Intern. Union for Conserv. Nature and Natural Resources, Warsaw (1960). Publ. in La Terre et la Vie, **108**(1): 116–129.

FRANZ, J. 1961. Biological control of pest insects in Europe. Ann. Rev. Entomol., **6**: 183–200.

FRINGS, H., AND M. FRINGS. 1957. Recorded calls of the eastern crow as attractants and repellents. J. Wildl. Mgmt., **21**: 91.

FRINGS, H., AND M. FRINGS. 1958. Uses of sounds by insects. Ann. Rev. Entomol., **3**: 87–106.

FRINGS, H., AND J. JUMBER. 1954. Preliminary studies on the use of a specific sound to repel starlings (*Sturnus vulgaris*) from objectionable roosts. Science, **119**: 318–319.

GANNON, N., AND J. H. BIGGER. 1958. The conversion of aldrin and heptachlor to their epoxides in soil. J. Econ. Entomol., **51**(1): 1–2.

GEDDES, S. R. (chairman). 1961. Vertical integration, family farm, agricultural chemicals, greenbelting, other. State of California, Assembly Interim Comm. Repts., **17**(9): 1–160.

GEIER, P. W., AND L. R. CLARK. 1961. An ecological approach to pest control. Proc. 8th Tech. Meeting, Intern. Union for Conserv. of Nature and Natural Resources, Warsaw (1960), p. 10–18. E. J. Brill, Leiden.

GENELLY, R. E., AND R. L. RUDD. 1956. Effects of DDT, toxaphene and dieldrin on pheasant reproduction. Auk, **73**(3): 529–539.

GEORGE, J. L. 1957. The pesticide problem. A brief review of present knowledge and suggestions for action. The Conservation Foundation, New York, 67 p.

———. 1958. The program to eradicate the imported fire ant. The Conservation Foundation, New York, 39 p.

———. 1959. Effects on fish and wildlife of chemical treatments of large areas. J. Forestry, **57**(4): 250–254.

———. 1960*a*. Some primary and secondary effects of herbicides on wildlife, p. 40–72. *In* Forestry symposium, Penn. State Univ., Aug. 30–31, 1960.

———. 1960*b*. Effects on wildlife of pesticide treatments of water areas, p. 94–102. *In* Proc. Symposium Coordinating Mosquito Control and Wildl. Mgmt., Washington, D.C., April 1–2, 1962.

——— (ed.). 1963. Pesticide-wildlife studies. A review of Fish and Wildlife Service investigations during 1961 and 1962. U.S. Dept. Interior, Fish Wildl. Serv., Circ. 167, 109 p.

GEORGE, J. L., AND R. T. MITCHELL. 1947. The effects of feeding DDT-treated insects to nestling birds. J. Econ. Entomol., **40**(6): 782–789.

GERSHON, S., AND F. H. SHAW. 1961. Psychiatric sequelae of chronic exposure to organophosphorus insecticides. Lancet, **1**: 1371–1374.

GINSBURG, J. M., AND J. P. REED. 1954. A survey of DDT-accumulation in soils in relation to different crops. J. Econ. Entomol., **47**(3): 467–474.

GRAHAM, E. H. 1944. Natural principles of land use. Oxford Univ. Press, New York. 274 p.

————. 1947. The land and wildlife. Oxford Univ. Press, New York. 232 p.

GRAHAM, R. J. 1960. Effects of forest insect spraying on trout and insects in some Montana streams, p. 62–65. *In* Biological problems in water pollution. Trans. 2nd Seminar Biol. Problems in Water Pollution (1959).

GRAHAM, R. J., AND D. O. SCOTT. 1959. Effects of an aerial application of DDT on fish and aquatic insects in Montana. Montana Fish Game Dept. and U.S. Forest Serv., 35 p.

GRATZ, N. G. 1959. The effect of dieldrin resistance on the biotic potential of the housefly in Liberia, p. 146–161. *In* Symposium on Pesticides, Brazzaville, Nov. 9–13, 1959. World Health Org.

GREEN, N., M. BEROZA, AND S. A. HALL. 1960. Recent developments in chemical attractants for insects. Advances Pest Control Research, **3**: 129–179.

GRESHAM, G. 1962. Pesticides vs. wildlife. Louisiana Conservationist, **14**(9, 10): 6–8.

GRZENDA, A. R., G. J. LAUER, AND H. P. NICHOLSON. 1962. Insecticide contamination in a farm pond. II. Biological effects. Trans. Amer. Fish. Soc., **91**(2): 217–222.

GUNN, D. L. 1960. The biological background of locust control. Ann. Rev. Entomol., **5**: 279–300.

GUNTHER, F. A., AND L. R. JEPPSON. 1960. Modern insecticides and world food production. John Wiley & Sons, New York. 284 p.

HAEUSSLER, G. J. 1952. Losses caused by insects, p. 141–146. *In* A. Stefferud (ed.), Insects—The yearbook of agriculture. U.S. Govt. Printing Office, Washington.

HALL, E. R. 1946. Mammals of Nevada. Univ. California Press. 710 p.

HARRINGTON, R. W., JR., AND W. L. BIDLINGMAYER. 1958. Effects of dieldrin on fishes and invertebrates of a salt marsh. J. Wildl. Mgmt., **22**(1): 76–82.

HARTENSTEIN, R. C. 1960. The effects of DDT and malathion upon forest soil microarthropods. J. Econ. Entomol., **53**(3): 357–362.

HAYES, W. J. 1959. DDT. The insecticide dichlorodiphenyltrichloroethane and its significance, p. 11–247. *In* S. W. Simmons (ed.), DDT insecticides. II. Human and veterinary medicine. Birkhäuser Verlag, Basel.

————. 1960. Pesticides in relation to public health. Ann. Rev. Entomol., **5**: 379–404.

HAYS, S. B., AND F. S. ARANT. 1960. Insecticidal baits for control of the fire ant, *Solenopsis saevissima richteri*. J. Econ. Entomol., **53**(2): 188–191.

HENDERSON, C., Q. H. PICKERING, AND C. M. TARZWELL. 1960. The toxicity of organic phosphorus and chlorinated hydrocarbons to fish, p. 76–88. *In* Biological problems in water pollution. Trans. 2nd Seminar Biol. Problems in Water Pollution (1959).

HENDERSON, W. W. 1931. Crickets and grasshoppers in Utah. Utah Agr. Expt. Sta., Circ. 96, 38 p.

HENSLEY, S. D., W. H. LONG, L. R. RODDY, W. J. McCORMICK, AND E. J. CONCIENNE. 1961. Effects of insecticides on the predaceous arthropod

fauna of Louisiana sugar cane fields. J. Econ. Entomol., 54(1): 146–149.
HERALD, E. S. 1949. Effects of DDT-oil solutions upon amphibians and reptiles. Herpetologica, 5: 117–120.
HESELTINE, N. 1960. The Quelea bird—a problem in international control. World Crops, 12(6): 209–212.
HESS, A. D. 1956. Public health importance of insect problems. Soap Chem. Specialties, 32(12): 191.
HICKEY, J. J. 1961. Some effects of insecticides on terrestrial bird life. *In* Rept. Subcomm. Relation of Chemicals to Forestry and Wildlife for the State of Wisconsin (Wis. Governor's Spec. Comm. Chem. Health Hazards). 55 p.
HICKEY, J. J., AND L. B. HUNT. 1960. Initial songbird mortality following a Dutch elm disease control program. J. Wildl. Mgmt., 24(3): 259–265.
HOFFMANN, C. H., AND E. P. MERKEL. 1948. Fluctuations in insect populations associated with aerial applications of DDT for forests. J. Econ. Entomol., 41(3): 464–473.
HOFFMANN, C. H., AND E. W. SURBER. 1949a. Effects of feeding DDT-sprayed insects to freshwater fish. U.S. Fish Wildl. Serv., Spec. Sci. Rept. (Fish.) No. 4, 9 p.
HOFFMANN, C. H., AND E. W. SURBER. 1949b. Effects of an aerial application of DDT on fish and fish-food organisms in two Pennsylvania watersheds. Progr. Fish-Cult., 11(4): 203–211.
HOFFMANN, C. H., K. H. TOWNES, H. H. SWIFT, AND R. I. SAILER. 1949. Field studies on the effects of airplane applications of DDT on forest invertebrates. Ecol. Monogr., 19: 1–46.
HOLLING, C. S. 1961. Principles of insect predation. Ann. Rev. Entomol., 6: 163–182.
HOOPER, A. D., AND J. M. HESTER. 1955. A study of thirty-six streams that were polluted by agricultural insecticides in 1950. Alabama Dept. Conserv., Fish. Rept., No. 12, 13 p.
HOWARD, L. O., AND W. F. FISKE. 1911. The importation into the United States of the parasites of the gipsy moth and the brown-tailed moth. U.S. Dept. Agr., Entomol. Bull. No. 91, 312 p.
HOWARD, W. E., B. L. KAY, J. E. STREET, AND C. F. WALKER. 1956. Range rodent control by plane. Calif. Agr., 10: 8–9.
HUFFAKER, C. B., AND C. E. KENNETT. 1956. Experimental studies on predation: Predation and cyclamen-mite populations on strawberries in California. Hilgardia, 26(4): 191–222.
HUNT, E. G., AND A. I. BISCHOFF. 1960. Inimical effects on wildlife of periodic DDD applications to Clear Lake. Calif. Fish Game, 46(1): 91–106.
HUNT, L. B. 1960. Songbird breeding populations in DDT-sprayed Dutch elm disease communities. J. Wildl. Mgmt., 24(2): 139–146.
HURLBUTT, H. W. 1958. A study of soil-inhabiting mites from Connecticut apple orchards. J. Econ. Entomol., 51(6): 767–772.
IDE, F. P. 1956. Effect of forest spraying with DDT on aquatic insects of salmon streams. Trans. Amer. Fish. Soc., 86: 208–219.

JACOB, F. H. 1958. Some modern problems in pest control. Sci. Progr., 46(181): 30–45.

JAMESON, E. W., JR. 1952. Food of deer mice, *Peromyscus maniculatus* and *P. boylei*, in the northern Sierra Nevada, California. J. Mammal., 33(1): 50–60.

———. 1958. The cost and effectiveness of controlling *Microtus* by zinc phosphide. J. Wildl. Mgmt., 22(1): 100–103.

JANZEN, D. H. 1961. The problem of nuisance and crop-depredating birds in the United States. Proc. 8th Tech. Meeting, Intern. Union for Conserv. of Nature and Natural Resources, Warsaw (1960), p. 79–84. E. J. Brill, Leiden.

JOHNSTON, D. W., AND E. P. ODUM. 1956. Breeding bird populations in relation to plant succession on the Piedmont of Georgia. Ecology 37(1): 50–62.

KALMBACH, E. R., AND J. F. WELCH. 1946. Colored rodent baits and their value in safeguarding birds. J. Wildl. Mgmt., 10(4): 353–360.

KEENLEYSIDE, M. H. A. 1959. Effects of spruce budworm control on salmon and other fishes in New Brunswick. Can. Fish Cult., No. 24: 17–22.

KEITH, J. O., R. M. HANSEN, AND A. L. WARD. 1959. Effect of 2,4-D on abundance and foods of pocket gophers. J. Wildl. Mgmt., 23(2): 137–145.

KENDEIGH, S. C. 1947. Bird population studies in the coniferous forest biome during a spruce budworm outbreak. Dept. Lands Forests (Canada), Biol. Bull. No. 1, 100 p.

KENNEDY, J. F. 1963. Use of pesticides. Rept. President's Sci. Advisory Comm., Washington, D.C., May 15, 25 p.

KERSWILL, C. J. 1958. Effects of DDT spraying in New Brunswick on future runs of adult salmon. Atlantic Advocate, 48(8): 65–68.

KLEINMAN, G. D., I. WEST, AND M. S. AUGUSTINE. 1960. Occupational diseases in California attributed to pesticides and agricultural chemicals. Arch. Environmental Health, 1(2): 118–124.

KOFORD, C. B. 1958. Prairie dogs, whitefaces, and blue grama. Wildl. Soc., Wildl. Monogr. No. 3, 78 p.

KUENEN, D. J. 1961. The ecological effects of chemical and biological control of undesirable plants and insects. Proc. 8th Tech. Meeting, Intern. Union for Conserv. of Nature and Natural Resources, Warsaw (1960), p. 1–9. E. J. Brill, Leiden.

LAIRD, M. 1958. A secondary effect of DDT upon the aquatic microflora and microfauna. Can. J. Microbiol., 4: 445–452.

———. 1959. Biological solutions to problems arising from the use of modern insecticides in the field of public health. Acta Trop., 16(4): 331–355.

LANGFORD, R. R. 1949. The effect of DDT on freshwater fishes, p. 19–38. *In* Forest spraying and some effects of DDT. Dept. Lands Forests (Canada), Biol. Bull. No. 2.

LATHAM, R. M. 1952. The predator question. The ecology and economics of predator management. Penn. Game Comm., Game News, Spec. Issue No. 5, 96 p.

LAUG, E. P., C. S. PRICKETT, AND F. M. KUNZE. 1950. Survey analyses of human milk and fat for DDT content. Federation Proc., 9: 294–295.

LAWSON, F. R., R. L. RABB, F. E. GUTHRIE, AND T. G. BOWERY. 1961. Studies of an integrated control system for hornworms on tobacco. J. Econ. Entomol., 54(1): 93–97.

LEMMON, A. B. 1957. Agricultural chemicals. Economic facts, laws, and regulations. A.M.A. Arch. Ind. Health, 16(6): 326–329.

LEOPOLD, A. 1933. Game management. Chas. Scribner's Sons, New York. 481 p.

————. 1949. A Sand County almanac. Oxford Univ. Press, New York. 226 p.

LICHTENSTEIN, E. P. 1957. DDT accumulation in mid-western orchard and crop soils treated since 1945. J. Econ. Entomol., 50(5): 545–547.

LICHTENSTEIN, E. P., AND J. B. POLIVKA. 1959. Persistence of some chlorinated hydrocarbon insecticides in turf soils. J. Econ. Entomol., 52(2): 289–293.

LICHTENSTEIN, E. P., AND K. R. SCHULTZ. 1959a. Breakdown of lindane and aldrin in soils. J. Econ. Entomol., 52(1): 118–124.

LICHTENSTEIN, E. P., AND K. R. SCHULTZ. 1959b. Persistence of some chlorinated hydrocarbon insecticides as influenced by soil types, rate of application and temperature. J. Econ. Entomol., 52(1): 124–131.

LILLY, J. H. 1956. Soil insects and their control. Ann. Rev. Entomol., 1: 203–222.

LINDQUIST, A. W., AND A. R. ROTH. 1950. Effect of dichlorodiphenyl dichloroethane on larvae of the Clear Lake gnat in California. J. Econ. Entomol., 43(3): 328–332.

LOGIER, E. B. S. 1949. Effect of DDT on amphibians and reptiles, p. 49–56. *In* Forest spraying and some effects of DDT. Dept. Lands Forests (Canada), Biol. Bull. No. 2.

LONG, W. H., E. A. CANCIENNE, E. J. CANCIENNE, R. N. DOPSON, AND L. D. NEWSOM. 1958. Fire ant eradication program increases damage by the sugar cane borer. Sugar Bull., 37(5): 62–63.

LONGHURST, W. M., A. S. LEOPOLD, AND R. F. DASMANN. 1952. A survey of California deer herds. Calif. Dept. Fish Game, Game Bull. No. 6, 136 p.

LOOSANOFF, V. L. 1960. Some effects of pesticides on marine arthropods and molluscs, p. 89–93. *In* Biological problems in water pollution. Trans. 2nd Seminar Biol. Problems in Water Pollution (1959).

McCORKLE, C. O. 1959. When will farm chemicals come into their own? Farm Chem., Aug.: 28–31.

MACFADYEN, A. 1957. Animal ecology—aims and methods. Pitman and Sons, London. 264 p.

MACPHEE, A. W., D. CHISHOLM, AND C. R. MACEACHERN. 1960. The persistence of certain pesticides in the soil and their effect on crop yields. Can. J. Soil Sci., **40**: 59–62.

MARTIN, H. 1957a. Guide to the chemicals used in crop protection. 3rd ed. Can. Dept. Agr., Ottawa. 306 p. and suppl.

————. 1957b. Chemical aspects of ecology in relation to agriculture. Can. Dept. Agr., Publ. 1015, 96 p.

MEANLEY, B., W. F. MANN, JR., AND H. J. DERR. 1956. New bird repellents for long leaf (pine) seed. Southern Forest Expt. Sta., New Orleans, Southern Forest Notes, **105**: 1–2.

MEHNER, J. F., AND G. J. WALLACE. 1959. Robin populations and insecticides. Atlantic Naturalist, **14**(1): 4–9.

METCALF, C. L., W. P. FLINT, AND R. L. METCALF. 1951. Destructive and useful insects—their habits and control. 3rd ed. McGraw-Hill, New York. 1071 p.

METCALF, R. L. 1955. Organic insecticides. Their chemistry and mode of action. Interscience Publ., New York. 392 p.

MINISTRY OF AGRICULTURE, FISHERIES, AND FOOD. 1961. Sixth report from the Estimates Committee together with the minutes of evidence taken before Sub-Committee D and appendices. House of Commons (Great Britain), London. 312 p.

MOHR, J. 1959. Influences of the poisoning program on wildlife, p. 27–34. *In* The Oregon meadow mouse irruption of 1957–1958. Fed. Coop. Extension Serv., Oregon State Coll.

MOORE, A. W., AND E. H. REID. 1951. The Dalles pocket gopher and its influence on forage production of Oregon mountain meadows. U.S. Dept. Agr., Circ. 884, 36 p.

MOORE, N. W. 1962. Toxic chemicals and birds: The ecological background to conservation problems. British Birds, **55**: 428–435.

MOORE, N. W., AND D. A. RATCLIFFE. 1962. Chlorinated hydrocarbon residues in the egg of a peregrine falcon (*Falco peregrinus*) from Perthshire. Bird Study, **9**(4): 242–244.

MORRIS, R. F., W. F. CHESHIRE, C. A. MILLER, AND D. G. MOTT. 1958. The numerical response of avian and mammalian predators during a gradation of the spruce budworm. Ecology, **39**(3): 487–494.

MRAK, E. M. (chairman). 1960. Report on agricultural chemicals and recommendations for public policy. Gov. Edmund G. Brown's Special Committee on Public Policy Regarding Agricultural Chemicals. Sacramento, Calif., 35 p.

MURIE, A. 1944. The wolves of Mount McKinley. U.S. Dept. Interior, Fauna of the National Parks, No. 5. 238 p.

NAGEL, W. O., F. W. SAMPSON, AND A. BROHN. 1955. Predator control—why and how. Missouri Conserv. Comm., 32 p.

NATIONAL AGRICULTURAL CHEMICALS ASSOCIATION. 1957. Economic benefits of pesticides. Spec. Issue, Natl. Agr. Chem. Assoc. News, **15**(3).

NATIONAL RESEARCH COUNCIL. 1952. Conference on insecticide resistance and insect physiology. Natl. Acad. Sci.–Natl. Research Council Publ. 219, 99 p.

———. 1956. Safe use of pesticides in food production. A report by the Food Protection Committee of the Food and Nutrition Board. Publ. 470, 16 p.

———. 1962. Reports of the Committee on Pest Control and Wildlife Relationships. I. Evaluation of pesticide-wildlife problems, 28 p. II. Policy and procedures for pest control, 53 p. Natl. Acad. Sci.–Natl. Research Council Publ. 920A-B.

NEFF, J. A., AND B. MEANLEY. 1957. Blackbirds and the Arkansas rice crop. Arkansas Agr. Expt. Sta., Bull. 584, 89 p.

NEGHERBON, W. O. 1959. Handbook of toxicology. Vol. III: Insecticides. W. B. Saunders Co., Philadelphia and London. 854 p.

NEW ZEALAND FOREST SERVICE. 1958. Proceedings of the meeting to discuss the noxious animal problem in New Zealand. Wellington. 133 p. and appendices.

NICHOLSON, A. J. 1939. Indirect effects of spray practices on pest populations. Verhandl. Intern. Entomol. 7th Kongr., Berlin (1938), 4: 3022–3028.

NICHOLSON, H. P. 1962. Pesticide pollution studies on the southeastern states. *In* Biological problems in water pollution. Trans. 3rd Seminar Biol. Problems in Water Pollution, 1962. 11 p. in preprint.

NICHOLSON, H. P., H. J. WEBB, G. J. LAUER, R. E. O'BRIEN, A. R. GRZENDA, AND D. W. SHANKLIN. 1962. Insecticide contamination in a farm pond. I. Origin and duration. Trans. Amer. Fish. Soc., 91(2): 213–217.

NIERING, W. A. 1958. Principles of sound right-of-way vegetation management. Econ. Botany, 12(2): 140–144.

NORRIS, J. J. 1950. Effect of rodents, rabbits, and cattle on two vegetation types in semidesert range land. New Mexico Agr. Expt. Sta., Bull. 353, 23 p.

ODUM, E. P. 1959. Fundamentals of ecology. 2nd ed. W. B. Saunders, Philadelphia. 546 p.

ONEY, J. 1951. Fall food habits of the clapper rail in Georgia. J. Wildl. Mgmt., 15(1): 106–107.

PARKER, J. R. 1942. Annual insect damage appraisal. J. Econ. Entomol., 35(1): 1–10.

PICKETT, A. D. 1961. The ecological consequences of chemical control practices on arthropod populations in apple orchards in Nova Scotia, Canada. Proc. 8th Tech. Meeting, Intern. Union for Conserv. of Nature and Natural Resources, Warsaw (1960), p. 19–24. E. J. Brill, Leiden.

PICKETT, A. D., W. L. PUTNAM, AND E. J. LeROUX. 1958. Progress in harmonizing biological and chemical control of orchard pests in eastern Canada. Proc. X Intern. Congr. Entomol., 3: 169–174.

PILLMORE, R. E. 1961. Pesticide investigations of the 1960 mortality of fish-eating birds on Klamath Basin wildlife refuges. U.S. Fish Wildl. Serv., Wildl. Research Lab., 12 p.

PIMENTEL, D. 1961. An ecological approach to the insecticide problem. J. Econ. Entomol., 54(1): 108–114.

PITELKA, F. A. 1957. Some aspects of population structure in the short-term cycle of the brown lemming in Northern Alaska. Cold Spring Harbor Symposia on Quant. Biol., 22: 237–251.

PITELKA, F. A., P. Q. TOMICH, AND G. W. TREICHEL. 1955. Ecological relations of jaegers and owls as lemming predators near Barrow, Alaska. Ecol. Monogr., 25(3): 85–117.

POPHAM, W. L., AND D. G. HALL. 1958. Insect eradication programs. Ann. Rev. Entomol., 3: 335–354.

POZNANIN, L. P. (ed.). 1956. Ways and means of using birds in combating noxious insects. Ministry of Agr., U.S.S.R. Originally in Russian; transl. by the Israel Program for Scientific Translation (1960). 138 p.

PRESNALL, C. C. 1949. Policies and philosophies of predator and rodent control. Trans. 14th N. Amer. Wildl. Conf., p. 586–592.

RADELEFF, R. D., W. J. NICKERSON, AND R. W. WELLS. 1960. Acute toxic effects upon livestock and meat and milk residues of dieldrin. J. Econ. Entomol., 53(3): 425–429.

REITZ, L. P. (ed.). 1960. Biological and chemical control of plant and animal pests. Amer. Assoc. Advance. Sci., Publ. 61, 273 p.

RIPPER, W. E. 1956. Effect of pesticides on balance of arthropod populations. Ann. Rev. Entomol., 1: 403–438.

ROBBINS, C. S., P. F. SPRINGER, AND C. G. WEBSTER. 1951. Effects of five-year DDT application on breeding bird population. J. Wildl. Mgmt., 15(2): 213–216.

ROBINSON, W. B. 1953. Coyote control with compound 1080 stations in national forests. J. Forestry, 51(12): 880–885.

RODRIGUEZ, J. G. 1960. Nutrition of the host and reaction to pests. In Biological and chemical control of plant and animal pests. Amer. Assoc. Advance. Sci., Publ. 61: 149–167.

RUDD, R. L. 1958. The indirect effects of chemicals in nature. Proc. 54th Ann. Conv. Natl. Audubon Soc., New York, p. 12–16.

———. 1959. Pesticides: The *real* peril. The Nation, 189: 399–401.

———. 1961. The ecological consequences of chemicals in pest control, particularly as regards their effects on mammals. Proc. 8th Tech. Meeting, Intern. Union for Conserv. of Nature and Natural Resources, Warsaw (1960), p. 38–47. E. J. Brill, Leiden.

RUDD, R. L., AND R. E. GENELLY. 1955. Avian mortality from DDT in California rice fields. Condor, 57(2): 117–118.

RUDD, R. L., AND R. E. GENELLY. 1956. Pesticides: Their use and toxicity in relation to wildlife. Calif. Dept. Fish Game, Game Bull. No. 7, 209 p.

RUDEBECK, G. 1950, 1951. The choice of prey and modes of hunting of predatory birds with special reference to their selective effects. Oikos, 2: 65–88; 3: 200–231.

SAMPSON, A. W. 1952. Range management, principles and practices. John Wiley & Sons, New York. 570 p.

SATCHELL, J. E. 1955. The effects of BHC, DDT and parathion on soil fauna. Soils and Fertilizer, **18**: 279–285.

SCOTT, T. G. 1943. Some food coactions of the Northern Plains red fox. Ecol. Monogr., **13**(4): 427–479.

SCOTT, T. G., Y. L. WILLIS, AND J. A. ELLIS. 1959. Some effects of a field application of dieldrin on wildlife. J. Wildl. Mgmt., **23**(4): 409–427.

SHEALS, J. G. 1955. The effects of DDT and BHC on soil Collembola and Acarina, p. 241–252. *In* D. K. McE. Kevan (ed.), Soil Zoology. Butterworth, London.

SHELLHAMMER, H. S. 1961. An ethological and neurochemical analysis of facilitation in wild mice. Univ. Calif., Davis. Unpubl. doctoral thesis.

SIMMONS, S. W. 1957. The current status of insecticide resistance. Industry and Tropical Health III, Proc. 3rd Conf. Ind. Trop. Health, Harvard, p. 34–41.

SMELSER, M. 1959. Uncontrollable "control." Nature Mag., **52**(1): 33–40, 50.

SMITH, K. D., AND G. B. POPOV. 1953. On birds attacking desert locust swarms in Eritrea. Entomology, **86**: 3–7.

SMITH, R. E. 1958. Natural history of the prairie dog in Kansas. Univ. Kansas, Museum Natural Hist., Misc. Publ. 16, 36 p.

SMITH, R. F., AND K. S. HAGEN. 1959. Impact of commercial insecticide treatments. Hilgardia, **29**(2): 131–154.

SOLOMON, M. E. 1949. The natural control of animal populations. J. Animal Ecol., **18**(1): 1–35.

SOPER, J. D. 1941. History, range, and home life of the northern bison. Ecol. Monogr., **11**(4): 347–412.

SOUTHEASTERN ASSOCIATION OF GAME AND FISH COMMISSIONERS. 1958. Proceedings symposium. The fire ant eradication program and how it affects wildlife. 12th Ann. Conf., 34 p.

SPECHT, H. B., AND C. D. DONDALE. 1960. Spider populations in New Jersey apple orchards. J. Econ. Entomol., **53**(5): 810–814.

SPECTOR, W. S. (ed.). 1956. Handbook of toxicology. Vol. I. W. B. Saunders Co., Philadelphia. 408 p.

SPRINGER, P. F., AND J. R. WEBSTER. 1949. Effects of DDT on salt marsh wildlife. U.S. Fish Wildl. Serv., Spec. Sci. Rept. No. 10, 23 p.

STEARNS, L. A., E. E. LYNCH, B. KRAFCHICK, AND C. W. WOODMANSEE. 1947. Report of the use of DDT and Paris green on muskrat marshes. Proc. New Jersey Mosquito Exterm. Assoc., 34th Ann. Meeting, p. 82–95.

STEFFERUD, A. (ed.). 1952. Insects—The yearbook of agriculture, 1952. U.S. Govt. Printing Office, Washington. 780 p. and appendices.

STEINIGER, F. 1952. Nagetierbekämpfung und Sekundärvergiftungen bei Raubvögeln und Eulen. Ornithol. Mitt., **4**: 36–39.

STERN, V. M., R. F. SMITH, R. VAN DEN BOSCH, AND K. S. HAGEN. 1959. The integrated control concept. Hilgardia, **29**(2): 81–101.

STERN, V. M., AND R. VAN DEN BOSCH. 1959. Field experiments on the effects of insecticides. Hilgardia, **29**(2): 103–130.

STEWART, C. P., AND A. STOLMAN. 1961. Toxicology—mechanisms and analytical methods. Vol. II. Academic Press, New York. 921 p.

STICKEL, L. F. 1951. Wood mouse and box turtle populations in an area treated annually with DDT for five years. J. Wildl. Mgmt., 15(2): 161–164.

STRICKLAND, A. H. 1957. Cotton pests and insect ecology. Empire Cotton Growing Rev., 34(4): 1–14.

TARZWELL, C. M. 1950. Effects of DDT mosquito larviciding on wildlife. V. Effects on fishes of the routine manual and airplane application of DDT and other mosquito larvicides. Public Health Repts., 65(8): 231–255.

———. 1959. Pollutional effects of organic insecticides. Trans. 24th N. Amer. Wildl. Conf., p. 132–142.

TAYLOR, W. P., C. T. VORHIES, AND P. B. LISTER. 1935. The relation of jack rabbits to grazing in southern Arizona. J. Forestry, 33(5): 490–498.

TEVIS, L. 1956. Behavior of a population of forest-mice when subjected to poison. J. Mammal., 37(3): 358–370.

THOMAS, W. L. (ed.). 1956. Man's role in changing the face of the earth. Univ. Chicago Press, Chicago. 1193 p.

THOMPSON, H. V. 1953. The use of repellents for preventing mammal and bird damage to trees and seed. Forestry Abstr., 14(2): 129–136.

THORSTEINSON, A. J. 1960. Host selection in phytophagous insects. Ann. Rev. Entomol., 5: 193–218.

TINBERGEN, L. 1946. De Sperwer als roofvijand van Zangvogels. Ardea, 34: 1–213.

———. 1949. Bosvogels en insecten. Ned. Bosbouw-Tijdschr., 4: 91–105.

TODD, F. E., AND S. E. MCGREGOR. 1960. The use of honeybees in the production of crops. Ann. Rev. Entomol., 5: 265–278.

TREHERNE, R. C., AND E. R. BUCKNELL. 1924. Grasshoppers of British Columbia. Can. Dept. Agr., Bull. 39 (n.s.), 47 p.

U.S. DEPARTMENT OF AGRICULTURE. 1948———. Notices of judgment under the Federal Insecticide, Fungicide, and Rodenticide Act. Plant Pest Control Div., occasional.

———. 1955. Agricultural statistics. 777 p.

U.S. FISH AND WILDLIFE SERVICE. 1962. Effects of pesticides on fish and wildlife: A review of investigations during 1960. Fish and Wildl. Serv., Circ. 143, 52 p.

U.S. FOREST SERVICE. 1958. Timber resources for America's future. Forest Resource Rept. No. 14. 713 p.

U.S. PUBLIC HEALTH SERVICE. 1961. Pollution-caused fish kills in 1960. U.S. Dept. Health, Educ., Welfare, Public Health Serv. Publ. 847, 20 p.

UVAROV, B. P. 1957. The aridity factor in the ecology of locust and grasshoppers of the old world. UNESCO, Arid Zone Research, 8: 164–198.

VAN DEN BOSCH, R., AND V. M. STERN. 1962. The integration of chemical and biological control of arthropod pests. Ann. Rev. Entomol., 7: 367–386.

VINSON, S. B., C. E. BOYD, AND D. E. FERGUSON. 1962. Resistance to DDT in the mosquito fish, *Gambusia affinis*. Science, **139**(3551): 217–218.

VORHIES, C. T. 1936. Rodents and conservation. Nature Mag., **28**: 363–365, 379.

VOÛTE, A. D. 1946. Regulation of the density of the insect-populations in virgin-forests and cultivated woods. Arch. Néerl. Zool., **7**: 435–470.

WALKER, K. C., M. B. GOETTE, AND G. S. Batchelor. 1954. Dichlorodiphenyl-trichloroethane and dichlorodiphenyldichloroethylene content of prepared meals. J. Agr. Food Chem., **2**(20): 1034–1037.

WALLACE, G. J., W. P. NICKELL, AND R. F. BERNARD. 1961. Bird mortality in the Dutch elm disease program in Michigan. Cranbrook Inst. Sci., Bull. 41, 44 p.

WEBB, F. E. 1960. Aerial forest spraying in Canada in relation to effects on aquatic life, p. 66–70. *In* Biological problems in water pollution. Trans. 2nd Seminar on Biol. Problems in Water Pollution (1959).

WEISS, C. M. 1961. Physiological effect of organic phosphorus insecticides on several species of fish. Trans. Amer. Fish. Soc., **90**(2): 143–152.

WELCH, J. F. 1954. A review of chemical repellents for rodents. J. Agr. Food Chem. **2**(3): 142–149.

WELLINGTON, W. G. 1960. Qualitative changes in natural populations during changes in abundance. Can. J. Zool., **38**: 289–314.

WESTHOFF, V., AND P. ZONDERWIJK. 1961. The effects of herbicides on the wild flora and vegetation in the Netherlands. Proc. 8th Tech. Meeting, Intern. Union for Conserv. of Nature and Natural Resources, Warsaw (1960), p. 69–78. E. J. Brill, Leiden.

WIGGLESWORTH, V. B. 1945. DDT and the balance of nature. Atlantic Monthly, **176**: 107–113.

WILSON, E. O., AND W. L. BROWN, JR. 1958. Recent changes in the introduced populations of the fire ant. Evolution, **12**: 211–218.

WOOTTEN, J. F. 1962. Methoxychlor: Safe and effective substitute for DDT in controlling Dutch elm disease. Central States Forest Expt. Sta., Station Note No. 156, 3 p.

WORRELL, A. C. 1960. Pests, pesticides, and people. Amer. Forests (July, 1960), reprinted by The Conservation Foundation, New York, 41 p.

WRAGG, L. E. 1954. The effect of DDT and oil on muskrats. Can. Field-Naturalist, **68**(1): 11–13.

WRIGHT, B. S. 1960. Woodcock reproduction in DDT-sprayed areas of New Brunswick. J. Wildl. Mgmt., **24**(4): 419–420.

YORK, G. T. 1949. Grasshopper populations reduced by gulls. J. Econ. Entomol., **42**(5): 837–838.

YOUNG, L. A., AND H. P. NICHOLSON. 1951. Stream pollution resulting from the use of organic insecticides. Progr. Fish-Cult., **13**: 193–198.

ZINNSER, H. 1934. Rats, lice, and history. The Cornwall Press, New York.

INDEX

Agriculture: crop losses from insects in, 44–49

Aldrin: in Japanese beetle control, 30; survival of residues in soil, 163; conversion to dieldrin, 166; cause of secondary poisoning, 242, 244

Amphibians, effects on: of DDT, 110–111; of heptachlor, 111

Animal weeds: defined, 193

Arsenical compounds: physiological action of, 15–16; in Japanese beetle control, 30; accumulation of, in soil, 163–164; adverse effects on plants from, in soil, 164

Artificial simplification. See Biotic simplification

Balance of nature: and pest abundance, 173; as ecological community concept, 187–188; imbalanced by pesticides, 268–269

Biological concentrators of pesticide residues: defined, 170, 248; exemplified, 248–267 passim. See also Residues; Delayed expression

Biological weeds: concept discussed, 193–194

Biotic simplification: contribution of pesticides toward, 176–177; naturally occurring, 189; induced by crop practices, 189–194

Birds: intentional control of, 117–118, 121–122; inadvertent effects on, by pesticides, direct poisoning – 123–124, reproductive changes – 137–140, 264–265, food shortage – 237, 239, indirect poisoning – 262–263, 248–267 passim

Bounty system: for animal control, 68

Buffer species: importance in substitution of foods, 226